Performance Improvement
Making It Happen

Performance Improvement
Making It Happen

DARRYL D. ENOS, Ph.D.

S^t_L

St. Lucie Press

Boca Raton • London
New York • Washington, D.C.

Library of Congress Cataloging-in-Publication Data

Enos, Darryl D.
    Performance improvement—making it happen / Darryl D. Enos
       p. cm.
    Includes bibliographical references and index.
    ISBN 1-57444-282-1 (alk. paper)
    1. Organizational effectiveness. 2. Performance.
    3. Teams in the workplace. I. Title.
    HD58.7.E4458 2000
    658.4—dc21
                                                 99-050959
                                                     CIP

No claim to original U.S. Government works
International Standard Book Number 1-57444-282-1
Library of Congress Card Number 99-050959
Printed in the United States of America   2 3 4 5 6 7 8 9 0
Printed on acid-free paper

Contents

Foreword

It has been my privilege to work with Darryl Enos for almost ten years. Beyond that, and perhaps more importantly, Darryl is a friend and is on my list of key people to go to when I seek counsel in dealing with my organizational, professional, and personal challenges and help in reaching my goals that are important to me at both the professional and personal level.

In the past, I have sometimes been somewhat bothered by the many books in which the author and the writer of the foreword appear to be involved in a kind of mutually supportive and self-congratulatory collusion. Darryl had some nice things to say about me in his preface and has now asked me to write this foreword and, as so often happens to us in life, I now find myself a participant in something I have been critical of in the past. I can only say that I appreciate Darryl's kind words and know that what I have to say about Darryl and this book could not be more sincere. So, once again I am forced to rethink a viewpoint and reframe an attitude. Not so bad though when you understand that the power of attitude and our willingness and ability to reframe attitudes are critical parts of what personal growth is about.

I want to share some observations with you concerning *Performance Improvement: Making It Happen*, but first I would like to share some thoughts about Darryl Enos. Helping you to appreciate Darryl I believe will help you to appreciate this very important book and hopefully encourage you to read it, ponder, and absorb its contents, and most important of all, use it to improve your own performance and the performance of individuals, teams, and organizations with which you are associated. Some key thoughts about Darryl:

- I could cite intelligence but it is much more important to note Darryl's insight and wisdom about people, situations, and life in general. I believe this insight is a function of both life experience and rational interpretation of the experience. Darryl's background and life experiences are rich and diverse. Like many of us he has experienced both great successes and great challenges. He has learned from both.
- Darryl combines the intellectual understanding and educational foundations one would expect from someone with a multidisciplinary doctorate in government, organizational theory, and social psychology

with diverse and deep, real world experience and the ability to seek, and usually find, practical and pragmatic solutions to the problems always associated with goals, pursuit, and achievement. In my own experience this combination of a strong educational background and intellectual understanding with the developed ability to operate successfully and make a real contribution in the real world is often all too rare. Clients have told me repeatedly that the two attributes they value most highly in employees are the ability to take action, to initiate, and the stamina to persist with follow through. Darryl consistently demonstrates both characteristics.

- Values are always a sensitive topic, but in the final analysis, they drive everything. All of us at Achievement Associates, Inc. are in the people development business. Without a sincere belief in the untapped potential of individuals, teams, and organizations and the conviction that this potential can be harvested, we could not do what we do as successfully as we do it. When Darryl joined Achievement Associates almost ten years ago, our early conversations focused almost entirely on values, vision, and goals, recognizing that shared values and vision and common goals are the foundation of any successful relationship or organization. Darryl has both the necessary belief in people and the conviction necessary to convert the energy of that belief into actions required.

- There is no more effective means of improving the performance of individuals, teams, and organizations than modeling the right kind of behaviors. Darryl knows the proven models of what works and continually applies himself to the challenge of modeling constructive, "people developing" behaviors. No one, despite good intentions, is as successful as they would like to be in modeling desirable behavior, but few people I know are as consistent as Darryl in providing the right kind of model.

- In Chapter 5 of *Performance Improvement: Making It Happen*, Darryl introduces the Model for Growth, probably the most significant and useful foundation model developed and used by our firm. The model suggests that there is a technology of people development that can be learned and successfully applied, sometimes with truly amazing results. Darryl came to our firm with a long list of accomplishments and a wide variety of positive attributes and developed skills that most people would consider exceptional. For ten years I have watched him use the six techniques for change embodied in our Model for Growth to further expand and even accelerate his own growth. And I know he is not done yet. This is the kind of modeling that I admire and appreciate and our clients have the right to expect.

To know Darryl is to know a lot about the book, but here are some observations about the book itself. Not being a stranger to the process of developing a book, and with a systems background and an appreciation for a "systems approach," I like the format and structure of *Performance Improvement: Making It Happen*. Part I, Performance Improvement: Getting it Started, provides necessary foundations in issues, philosophies, definitions, language, and models related to performance improvement and also moves the reader into the "how to" and "what needs to be done" to get started. The reader cannot legitimately fall back on the avoidance behavior crutch of "I can't get started until I finish the book and understand everything." The committed reader can get into action early on.

Part II, Performance Improvement: Taking Action, provides a wealth of tools and techniques aimed at affecting real change and positive improvement. This is the heart of the book when it comes to committing to and even carrying out actions for performance improvement.

One part of Darryl's book I particularly appreciate is Part III, Evaluating and Stabilizing Performance Improvement. Too often good ideas and the means for their implementation are presented with no attention to evaluating levels of success and sustaining, and even enhancing, the performance improvement process over time. I believe the primary ingredients of developing the performance of individuals, teams, and organization over time are belief, patience, and commitment. Ongoing measurement is also critical. Also implied is first a necessary tolerance for the frustration that occurs when progress seems too slow and second the kind of resilience and tough mindedness that only core values and long term commitment can engender and sustain. The use of cases is appropriate, balanced, and offers insight born of practical experience. I admire Darryl's candor in describing not only successes, but also what went wrong or "not so right." All real progress involves periodic setbacks. In addition to the cases, the suggested action steps for leaders are particularly useful.

It is recognized throughout the book that values, vision, and goals drive everything. They result in positive results and achievement. Unclear values, vision, and goals most often produce no results or even negative results. Darryl recognizes this.

The ancients understood that a common language is the beginning of real understanding and effective, efficient communication. This book does a great job of providing a host of models and language components to improve understanding and communication. The vast majority of these models have been proven effective by our day-to-day use of them in our work with Achievement Associates' clients. In general, *Performance Improvement: Making It Happen*, offers a wide variety of tools and techniques and virtually all of them are proven through our work at Achievement Associates or elsewhere. It is a cafeteria line; review what is available, and take and use what you need.

To you, Darryl — as you know I am presently involved in developing an in-depth training and development program called "Mentoring for Achievement." The research that I have done on mentoring as practiced since the beginning of recorded history (and perhaps even before) has been interesting and revealing. One thing is very obvious – mentoring has always been a two-way street. If I have been your mentor, you have certainly been mine and I thank you. Congratulations on a great and very needed book.

To you, the reader — if you are committed to performance improvement for yourself, for other individuals, teams, or an entire organization, read this book. More importantly, seek improved understanding in its pages, and put into action what you learn. Good intentions alone are never enough. Commit to start the performance improvement process now and stick with it. The results for you, others, and your organization may well amaze you.

Eric Hoffer, the longshoreman poet, said:

> Those who invest themselves in becoming their best selves, and even more importantly, those who invest themselves in helping others become their best selves, are involved in the most important work possible on the face of the earth, they are helping to complete God's work.

Certainly there are issues of organizational success and bottom line results, no matter how they are measured, but in the final analysis, the work described by Eric Hoffer is what this book is about.

Mike Weaver
Hilton Head Island, SC
January 6, 2000
President, Achievement Associates, Inc.

Preface

Organizations are different from each other, and yet they are alike in many ways. Organizations of various types (large versus small, not-for-profit versus profit-driven, etc.) usually have similar challenges, problems, and opportunities that make performance improvement important to them. The specific details of their challenges, problems, and opportunities will be different between organizations doing different types of business. But many of the basic causes of excellence or deficiency in performance are the same across different organizations and different work teams.

Just as there are similarities in the causes of performance levels between different organizations, there are also some similarities between the causes of good performance in organizations and what causes it in teams. This means that there are a number of basic approaches that can improve performance in organizations and teams, regardless of who they are and what they do. Indeed, there are basic similarities in the numerous approaches for improving performance in organizations, teams, and the people within those groups.

This is not to say that organizations and teams have no significant differences from one to another. Clearly, organizations and teams are influenced, for example, by the personalities, attitudes, motivations, and values of their members which vary from one organization and team to another. They are also influenced by the types of work they do. But the perspective in this book is that all organizations and teams, no matter what business they are in, are in large part the creation of their team members and improving performance originates from their acceptance of the responsibility to do so. The characteristics and causes of effective performance in organizations and teams tend to be very similar regardless of the type of work performed.

There are, of course, differences among organizations and teams based on what services or products they offer. Retailers, route businesses, manufacturers, non-profit hospitals and universities have different products to develop and provide and different markets to serve. But all of them will perform better if they have a clearly defined direction, a strategic vision. Furthermore, the problems that occur when the strategic vision is unclear tend to be similar between these organizations despite the fact that they have different products and clients.

Leadership and organizational climate will also differ from one organization or team to another. That is why clear understanding of the leadership style and climate is important to understanding performance. But ineffective leadership produces similar results no matter what team or organization is involved: lack of motivation, poor retention of employees, and lackadaisical customer service, for example. So, there are basic similarities in causes of poor levels of performance.

One similarity, for example, is that no matter what team or organization needs performance improvement, part of the approach should normally include helping people learn to work together more effectively. There are techniques for doing this that have a significant impact. The variety comes in the areas of performance a team or organization needs to improve.

A second similarity between organizations and work teams is the critical role of leadership in defining, supporting, and "institutionalizing" performance improvement. Experience shows that more failures in performance improvement efforts occur because of lack of leadership dedication and involvement than for any other single reason. More often than not, the failure to get leadership sufficiently involved is the result of the irresponsibility of people in my profession: facilitators and consultants in performance improvement, management development, and organizational/team development.

This book is comprised of three parts. Part I, Performance Improvement: Getting It Started, begins with a discussion in Chapter 1 of why so many organizational leaders feel the need to improve performance and specific reasons why they get involved in these efforts. The importance of how and why organizational/team leaders start their discussions and considerations regarding performance improvement is described, and the major costs of a poor beginning are identified. Brief actual cases are used for illustration.

Chapter 2 takes the definition of performance presented in the first chapter and further defines it in the organizational and team context. Also discussed are general perspectives on identifying performance deficiencies or gaps.

Chapter 3 details a process for a collaborative approach between organizational leadership and performance improvement experts, whether internal or external, in deciding on performance improvement for the team or organization. The objectives and knowledge and skills required for each stage are identified. Here, we make the case for the critical role of diagnosis and what is required for diagnosis to succeed.

Chapter 4 focuses on goal setting, cascading goals, the related topics of "Key Performance Indicators," and the role of feedback in measuring and achieving performance.

Part II, comprised of Chapters 5 through 12, outlines a number of performance improvement interventions including approaches, concepts, tools, and techniques. Brief cases are used throughout these chapters to help illustrate major concepts and to demonstrate where and why performance improvement succeeded or failed. All of these cases are ones where I, and sometimes other colleagues, had direct involvement.

Part III, which includes Chapters 13 through 15, focuses primarily on approaches for making performance improvement permanent. One of the great frustrations for both leaders and those assisting leaders in building performance is that results are often only temporary. Greater attention to institutionalizing progress is needed in many cases.

There are two features to this book that should be extremely useful to readers, especially those who make decisions about performance for organizations or teams. Throughout the book, the brief presentation of actual cases is used to illustrate a point that is particularly critical. This will help those who learn best by seeing practical examples. Again, all of the cases are ones in which I was a facilitator or consultant along with colleagues. I have included cases where the results were not positive as well as those where performance improved. In a few cases I have not listed the name of the organization because of their policy on public relations. But the cases are real.

Secondly, I have included a list of suggested action steps for leaders at the end of each chapter. If the reader is interested in considering whether performance improvement is warranted in his/her organization or team, or what options are available for making performance better, the suggested action steps will be useful. These suggestions are somewhat generic, but will help readers think about the application of the ideas in the chapter to their situation.

My professional focus and motivation has always been toward application and use of performance improvement approaches and concepts in actual situations. As a result, I have constantly sought mentors who could help me learn basic processes, concepts, and models for understanding and improving performance. Many people have been generous in providing suggestions and collaboration in my professional development and in writing this book.

Mike Weaver, president of Achievement Associates, Inc., is the most recent and perhaps the most significant mentor I have had the good fortune to work with. Mike, a technically trained engineer who decided people were a more important focus than systems, developed or helped develop many of the concepts and models contained in this book. With Mike Weaver, and with others, I have tried to give appropriate credit for their creations throughout the pages of this book.

Tony Montebello, Ph.D., was my first mentor after I changed careers to move into performance development work full time. Tony helped me learn attitudes, behavior, tools, and techniques necessary for success in the difficult world of performance consulting. Old habits, learned through years of academic teaching and administration, died hard but did disappear over time.

Paul Sultan, Ph.D., now retired from Southern Illinois University, was my co-author in my first published book, *The Sociology of Health Care*. I learned a great deal from Paul about authoring a work that is directed toward helping those interested in performing better. I hope I have applied those lessons well in this book.

Many current and recent clients have been active in discussions with me regarding this book and cases that are contained in it. Most notable are Jane Wulf and Rodger Reiney (CEO) from Scottsdale Securities, Dave Vaughn of Corporate University Xchange, Inc., and Dave O'Keefe (CEO) of Landshire, Inc. Others with whom I have had a valued professional relationship include Ted Perryman, managing partner of the law firm discussed in this work, Rob Anderson and Sue Rubin of Tone's Brothers, Gary Rager of American Car Foundry, and Ben Ernst, President of Ranken College. Technical support from Kevin and Joanie is also greatly appreciated as well as advice from Mel Kohr.

Writing a book while continuing to work in the field about which you are writing has advantages and challenges. Many of the people listed above have helped me confirm or at times amend my perspective on what occurred in cases discussed in the following pages. Tom and Peggy Schroeder have helped me keep professional perspective while remaining valued, close friends.

My wife Bonnie has provided personal and professional support during the past few months without which this book would not exist. Our children, Tracy, Erika, and Derick, and our grandchildren, Lena and Tyrese, have been supportive, tolerant, and sources of fun and relaxation. I thank them all.

Darryl D. Enos

About the author

Since completing his Ph.D. at the Claremont Colleges in California, Darryl D. Enos has worked for more than three decades combining performance improvement efforts at hundreds of organizations with university-level teaching and administration. During that time, the organizations he assisted include Fortune 500 companies, mid-size organizations, and not-for-profit organizations in virtually all major categories of products and services.

Dr. Enos held full-time teaching and administrative positions at four universities including 5 years at Claremont Graduate School where he was an associate of Peter Drucker. During all of his academic career, Enos has worked with organizations as diverse as the California State Assembly Agricultural Committee, May Company Department Stores, Toyota USA, Texas Instruments, and Anheuser-Busch, Inc., as well as many smaller organizations with less name recognition. He has focused on developing and refining concepts, tools, and techniques for diagnosing and improving performance.

Enos has written a number of publications. In the late 1970s, he co-authored the *Sociology of Health Care*, a textbook written with Paul Sultan, Ph.D., now a retired professor from Southern Illinois University. Other work includes a report to the Markle Foundation on Spanish Language Only Television and a contribution to the California Joint Committee Master Plan for Higher Education.

Dr. Enos is currently a senior associate with Achievement Associates, Inc. (AAI) in St. Louis, Missouri where he has worked for more than nine years. This firm has served over 450 clients throughout the United States since its beginning in 1972. AAI focuses on strategic planning, leadership and management development, improving sales and customer service, effective hiring, and team building. Darryl's current clients include Anheuser-Busch College of Sales and Marketing, Scottsdale Securities, Inc., Tone's Integrated Ingredients in Iowa, Landshire, Inc., Ranken Technical College, and numerous other collegiate, corporate, and governmental organizations.

part one

Performance improvement:
Getting it started

chapter one

Performance improvement: An overview of today's realities

This book offers concepts, models, processes, tools and techniques for improving performance in organizations and work teams of all types. But any decision on performance improvement has to be based on agreement about what performance is and how to assess it in an organization or team. One of the principal concepts reiterated throughout these pages is the paramount importance of clear purpose and direction in efforts at performance improvement. What is it we are trying to improve? In short, what is performance?

A. Performance definitions

Many definitions of performance are suggested by organizational leaders, decision makers, and students of organizational management and performance. A large percentage concern financial performance: profitability, gross income, or margins. One limitation to this focus is that it tends to exclude organizations, primarily government and non-profit organizations, where funding comes from someone other than the customer or consumer. The principal limitation of this definition of performance, however, is that it ignores the fundamental question: What do we need to accomplish to make the profit, income, or margins? To paraphrase a concept often attributed to Peter Drucker: figure what business you're in and how to do it well, and the money will follow.

Generating income is important to success and survival, even in many non-profit organizations such as colleges and universities or cultural endeavors. But financial health as a sole definition of performance does not tell us what to do to be successful. Something else is needed. Some organizational leaders, therefore, talk about being the best in their business. A phrase offered by thoughtful CEOs is "world class in the business of ..." But the flaw with this definition of performance is the same as exclusive focus on profit or income as a way of defining performance: What do we need to do to become "world class"?

Some university textbooks, often written for adult students who are managers, seem to get closer to a useful definition. An outstanding book, *Organizational Development and Change* says that organizational development (O.D.) is oriented to improving organizational effectiveness. Thus, "an effective organization is able to solve it own problems and focus its attention and resources on achieving key goals."[1] This definition is a start at defining performance, but it assumes the goals are clearly defined in the organization. Many organizations, of all sizes and types of business, have few if any stated performance goals, and those that do, often keep them general and vague.

It is useful to define performance in the following way: "performance is the definition and progressive achievement of tangible, specific, measurable and personally meaningful goals."[2] A few key points listed below will help clarify the meaning of this definition.

1. Many organizations, teams and individuals, both at work and personally, do not have well-defined goals. It is common to find smaller organizations, perhaps 20 million dollars in gross income, with no goals other than a structured budget. Even the budget may be poorly defined. Larger organizations, Fortune 500 companies for example, often have vague and general goals or objectives at the corporate level and provide little specific direction to those charged with achieving these targets. More about that later.

2. Setting specific and measurable goals forces us to decide what we are really trying to achieve and how to assess success. Measurement in work situations, especially in areas other than finance or production, can be scary to managers and employees for reasons ranging from difficulty of defining "soft goals" to fear of failure. However, as we discuss in Chapter 4, goals can be created in a way that provides clear direction and increased chances of performance success.

3. The fact that goals should be personally meaningful is probably the most critical point in the above definition of performance. Goals should not just appear out of nowhere, nor does having goals guarantee commitment and motivation to achieve. To get commitment and motivation, organizations need to start with the following question, "Goals to do what"? The initial answer which can increase commitment is "Goals to accomplish the vision, the drive for success, and the desire for competence in those involved in the team or organization."

B. Why the powerfully felt need for performance improvement by leaders?

Efforts at performance improvement in organizations, teams, and with individual employees are everywhere. Training and development, only one of the major efforts at performance improvement, is estimated to have cost U.S. companies over 55 billion dollars in 1995.[3] A 1994 survey done by AMA/Deloitte & Touche demonstrated that 84% of the companies included in the study had change initiatives underway, 68% had formalized change management programs going, and close to half of the companies had three change initiatives in effect.[4]

The factors that led to the huge number of performance improvement initiatives in the mid-1990s have probably increased in intensity, meaning that the efforts at performance enhancement have undoubtedly increased in recent years. The list of factors creating this powerfully felt need varies with the person stating it, but the following list is better than most.

1. **Competition** is a main concern of businesses wanting to improve their performance, especially organizational decision makers facing intense and worldwide competition. A few years ago, I discussed a set of performance issues and some possible improvement efforts with a small construction company. The significant information on that company was as follows:

- Owner operated company in a small municipality within a larger city in the Midwest.
- Products and Services
    constructed steel beams as reinforcement for commercial
        builders
    delivered and helped install those beams
- Markets
    construction companies
    commercial architects
- Structure
    owner operated
    a total of approximately 20 employees
- Primary Issues
    intense competition and the need for a redefinition of their
        marketing approach
    some problems with leadership and management style

In discussion with this group of managers, it became clear that their primary concern centered on their new intense competition. Their competition came mostly from Pacific Rim companies that provided products to the central United States. The one world economy, which had a dramatic impact on the American auto industry, is with us to stay, from manufacturing, to services, to education, and even to high technology. Performance improvement is a requirement for survival in situations of intense competition.

2. An increase in **customer knowledge and demands** also drives the felt need for performance improvement among many organizational leaders.

This increase in customer knowledge and demands is fed by the increased ease of access to information in today's networked world and the intensity of competition where customers have so many options. In addition, the recent popularity of total quality programs taught at least some customers, and sometimes their suppliers, that quality means doing what is right for the customer at all costs.

3. **Rapid technology changes,** and a rapid and increasing pace of change generally, often lead organizational leaders to feel the need to make efforts aimed at performance improvement. This need takes many forms: "our customers expect us to have and use this technology;" "the competition is getting it and will be ahead of us;" or, "this technology will make our team more effective when we learn how to make the best use of it." In some situations, achieving the most recent technology becomes an end in itself, rather than a significant component in a broader performance improvement initiative. But whatever its use, major technology change often means training or restructuring to increase or maintain performance.

4. **Human resource needs and desires,** including people throughout the organization are often factors in motivation performance improvement. This includes a number of well-known human resources related trends: the desire of leadership to build shareholder value; the impact of downsizing and the need to do more with less; changing attitudes towards work (Generation X); the emphasis on the "knowledge worker;" shortage of labor supply which results in jobs often going unfilled and other employees needing to be more productive to cover the work.

5. Some experts on organizational behavior believe **that human beings have a powerful need to be competent.** Authors like Jay Hall, Ph.D., see that need to be competent as the basis for our progress historically. Hall states, "If excellence is our goal, we already have what it takes to get there. We are a people with both the desire and the ability to do good work! We are competent."[5] The need to feel and be competent is a motivator in many people. Others seem to have misplaced it or had it driven out of them. Still, when people are motivated to get better at what they do, it often leads to considerations of performance improvement initiatives.

6. Incredible and growing **knowledge availability,** ranging from information about customers or markets and market niches, competition, or knowledge about new technology or suppliers and their wares, often produces a strong drive to learn this knowledge and integrate it into performance. Because of the abundance and array of information and the constant growth in the supply of data, organizations attempt to make use of the information often at a frantic pace.

C. Limited success in performance improvement

While there are so many strong factors motivating the need for performance improvement in our organizations, there are both experience and research to support the conclusion that "most companies have achieved only modest success in change management."[6] One of the primary purposes of this book is to identify how and why so much of our effort at performance improvement either fails or experiences very limited success. A second purpose for writing this book is to demonstrate how performance improvement initiatives can be made more successful.

Success or failure in performance improvement efforts begins with the reasons why organizational decision makers decide to get involved in the first place. The early discussions and thinking of organizational decision makers about performance often set the direction and tone of what, if anything, is to be done. Sometimes performance improvement consultants, from within or outside the organization, are involved in early decisions and therefore have the opportunity to clarify the perspective and intentions of organizational/team leaders. Other times consultants are forced into the role of the helping hand and told what to do. If they accept that role, clarifying the motivations of organizational leaders in seeking performance improvement may be a lost opportunity. Whatever the process for early decisions about performance improvement, the motivation and commitment of the leadership should be clarified early to increase chances of success.

A number of years ago, our firm was involved in discussions concerning leadership development with a highly placed executive of a very large communications company outlined below.

Organization: A Fortune 500 company with worldwide reach

- Products and Services
    communications services
    communications equipment
- Markets
    households
    businesses of all sizes
- Structure
    regionalized structure with a good deal of local operational
        autonomy
- Primary Issues/Problems
    apparent need for improving management/leadership style
    vagueness about why the decision maker was interested

In initial conversations, the motivations of the leader of the group of managers were hard to uncover. Vague phrases pertaining to the need for improved leadership style, both from the primary decision maker and from others in his group involved in early discussions, were the primary data provided. The collaborative or "partnering" approach so valuable in performance improvement did not occur. The leader of the group continued to keep his motivations for management development a secret.[7]

During the performance improvement interventions, additional data collection about the management group continued, and some basic conclusions became clear. First, the top leader himself was a major part of the leadership problem, and others knew it. Secondly, his motivation for performance improvement was primarily to meet a commitment he made to his own manager about improving his team. In short, he had not really thought about what his group needed for performance improvement and had little interest in collaborating to determine needs or to help make his team better. His motivation was exclusively to satisfy his boss.

Throughout this book , there is the stated position that the commitment and involvement of leadership is the most critical element in whether or not performance efforts, when initiated, are successful. Support will be provided for that point of view a number of times. In the context of this discussion, however, one basic point is important. If the organization/team leader is not motivated by authentic interest in and commitment to performance of his/her group, and is not open minded about doing what is best for developing that group, performance improvement efforts have little chance to work. Being authentically interested in and committed to performance is usually essential to being open minded about what to do and committing to search for the best methods of improvement.

D. Specific reasons why leaders start performance improvement

Anyone reading the brief description of the above case describing the Fortune 500 communications company might conclude that we were lucky to have leadership support, whatever the leader's motivations. This gets us directly to the important issue of why organizational leaders get involved in performance improvement and the impact it has on success. Listed below are some of the primary motivations for focus on performance improvement that are frequently found among organizational/team decision makers.

1. *The leader said so* is a common reason why organizations start a structured program for improving their performance. Leadership support is absolutely necessary for performance improvement in any organization. But it is not sufficient. The leader's support must be such that he or she is dedicated to do everything necessary for making performance improvement happen. In the communications case cited above, the local leader's own boss wanted

him to work on management style and that was the primary motivation. The managers below the local leader got involved in the management development program because their local boss said it was required.

There are a number of negative consequences from engaging in performance improvement solely because the leaders said so. First, organizational leaders may or may not understand what is causing performance deficiencies, but they almost never know how to improve it. When acceptance of a performance program is based only on the "boss said so," the strong tendency is to not only accept the preliminary judgment that improvement is needed, but to accept preliminary decisions about the specific nature of the problem and what is needed to fix it.

Secondly, someone has to confront and clarify the leader's motivations to make sure dedication for performance improvement is sufficient to sustain it. Someone from inside or outside of the organization needs to help the leader make sure that there is sufficient dedication for making performance improvement happen. Finally, the perceptions of the leader regarding the team or organization should be considered but cannot be taken as the final word. As we discuss in Chapter 3, the diagnosis of the performance situation and what needs improvement is the most critical stage in efforts at developing a team or organization and should not depend solely on one person's point of view. When people of high position within the organization/team are motivated to make "something happen' with their group, the tendency is to accept their judgments about what to do without sufficient diagnosis of what is really going on in performance.

Confronting a top decision maker or CEO in any organization about his or her motivations and perceptions is risky business. It is risky whether the person doing the confronting is from inside or outside the organization. The key here is why and how the confrontation is conducted. Good intentions go a long way.

2. *There is money* for developing our organization/team is a second common reason why decision makers consider performance improvement. This reason for engaging in performance improvement is similar to *the leader said so*; it sounds good but has its dangers if diagnosis stops there. Developing organizations, teams, or individuals costs money, many billions of dollars each year in the United States alone. But being able to meet the practical necessity of paying for the efforts is only a start. The key step is to thoroughly diagnose the performance situation to identify problems, issues, and causes. This is discussed further in Chapter 3.

The reverse situation, *we have no money* also occurs frequently, and lack of money ends the decisions about performance improvement. Because of the difficulty of measuring return on investment for performance improvement efforts, these efforts are usually seen as cost centered, and are therefore a luxury to be forgotten when money is tight.

Organizations have budgets, and budgets are set up on a time limit, usually one year. It is tempting to fall into the "we better spend it or we will lose it" trap. The availability of money, like the support of the leader, is desirable but cannot be the sole or primary motivation for engaging in performance improvement.

3. *Others are doing it,* sometimes referred to as the fad motivation, is subtly powerful in decision making of organizational/team leaders and performance improvement facilitators. Sometimes fad as a reason for doing something comes from external consultants who have the "latest and greatest thing" that they want to sell.[8] Just as often, the ideas about what others are doing comes from managers within the organization and CEOs who talk to and influence each other in various meetings and encounters.

CEOs and top managers are often a close knit group within a geographic area or an industry. They spend a lot of time talking with each other at professional associations, during board of directors meetings at the local bank and in social gatherings. They are often a valuable source of ideas and information to each other. But they can also have an undesirable influence. Because managers share leadership responsibilities, they have empathy with each other and experience similar issues. Only infrequently, however, do they know how to diagnose or improve performance. That is even more true if the pre-emptive diagnosis is done during happy hour and done on the basis that "it is helping us and it may help you."

There is a special version of the fad, the idea that a performance improvement effort worked before for the organization/team involved, and therefore it should be done again. As is the case with other motivations for decision maker's involvement in performance improvement, the problem here is the tendency to jump to action and minimize diagnosis. Again, the result is insufficient understanding of the performance status, deficiency, causes, and possible solutions.

Dedication and active support from organizational leaders are critical to the success of performance improvement. Having the finances to support the program is also critical, and there is nothing inherently wrong with a performance improvement program being popular. The question is whether these motivations dominate the initiation of the program and result in inadequate diagnosis of the issues and opportunities. The specific intervention aimed at performance improvement, whether it is restructuring, strategic planning, a major training and development effort, or some other intervention, cannot succeed if it is aimed at the wrong target. Knowing where performance currently is, identifying deficiencies clearly and understanding the organizational factors, attitudes and behaviors that are barriers to that performance are critical. They are the essence of diagnosis.

We started this chapter by defining performance as the definition and progressive achievement of tangible, specific, measurable and personally meaningful goals. If the performance improvement efforts themselves lack defined, tangible, specific, measurable and meaningful goals, then those efforts at building organizations and teams are doomed to partial success at best and failure at worse. A brief actual case can help illustrate this principle.

- Background of the Organization
  small consumer services company operated by founders' sons
  owned by two brothers who were very concerned about the
  company's future
- Products and Services
  servicing and installation of HVAC systems
  maintenance contracts on HVAC systems
  higher technology air quality systems
- Markets
  households
  commercial buildings
- Primary Issues/Problems
  decline in gross income over recent years
  issues of economic decline in their geographic service area

In the original conversations with the owners of this 40 year old company, it became clear they had decided on the nature and solution to their problem. They believed they needed to improve their customer service so as to get their lost business back and keep what they still had. After lengthy discussions, including review of the way they had defined their customer groupings (markets), it became clear that the urgent need was to expand geographic markets, find ways to provide services further from the home office, and improve selling efforts. After a number of discussions, the open minded owners changed their opinions and agreed to efforts aimed at these strategic and selling performance needs. Customer service was not the urgent need. Providing better service to a market shrinking because of demographic shifts would not solve the basic problems.

The owners of this smaller company wanted to be successful, but they weren't sure how to define that or how to get there. Increased clarification of their vision of the company and what it should be like in the future was essential if they were to succeed. An early step was to define the specific goals of the performance improvement efforts. They were as follows: redefine expanded markets to be served; develop a strategy for selling effectively to penetrate those markets; and find approaches for distribution of the services to be provided to the expanded markets. Once this strategic vision was established, other efforts at performance improvement, including customer service for the newly defined markets, were initiated by the company.

E. A process approach: The first look

The main point of this case, discussed again later in more detail, is that these leaders needed to clarify their vision or expanded mental picture of the desired and intended future of the company they owned. But they hadn't succeeded in doing it alone. We all tend to benefit from discussions of important topics with others who are well intended.

Organizational leaders can benefit from open discussions about performance either with people working with them or with improvement experts. It is often very productive to use a collaborative process to define the leaders' vision and understanding of the needs of the organization.

When discussion with others is going to occur, the organization/team leader should bring dedication to candor and willingness to share both hard data about the organization and their impressions and tentative conclusions about current and future performance. The internal or external performance improvement expert should bring, or be quick at developing, models of effective performance for the area(s) of concern, as well as a great ability to listen and collect information about the situation. Much more about this in Chapter 3.

As the issues, needs and opportunities are clarified, possible interventions to improve performance will emerge based on this diagnosis. Many interventions are available, and are discussed throughout this book. But the most critical stage in performance improvement, perhaps surprisingly, is not the intervention, rather it is the diagnosis.

Organizational leaders facing the need to participate in diagnosis and possible performance improvement efforts have a tactical decision to make. That decision relates to the scope of the performance improvement effort. A simple distinction is between the micro level and the macro level. Micro level refers to the leader thinking the focus should be limited to developing his or her supervisory team, sales team, customer service group, or MIS department. Sometimes micro focus includes developing the performance of a single individual who is important to the organization.

Issues of performance do at times appear particularly critical in one area of the organization, and therefore focus on the micro level is appropriate. A challenge of micro focus that must be understood, however, is that organizations are systems, and working to improve performance in one area may have unforeseen and even negative consequences in another. For example, as experience has shown a number of times, improving production or sales in a manufacturing company may cause problems unless the logistics department responsible for supplies of raw material is made aware of the changing need for these materials. Also, improving supervisory knowledge and skills to increase the amount of on the job training they provide probably means those hiring new supervisors should clarify that this is expected of new supervisors as well.

In the best of all worlds, focusing on the macro level is more desirable because it offers the opportunity to move all facets of the organization ahead.

This does not mean that "we have to do everything at once," as many CEOs fear. Rather, it means that there is an organizational performance improvement plan which involves coordinated and timed efforts in each area of performance improvement where the need exists.

F. Summary

William James, famous American philosopher of the early 1900s, is credited with having said that the way we begin an endeavor is the most critical factor in its success. Effective diagnosis of the performance needs, deficiencies, and issues is essential to success in performance improvement. But effective diagnosis requires candor and good intentions on the part of organizational/team decisions makers.

Organizational decision makers and performance improvement consultants alike need to pay attention to the way they begin the important endeavors aimed at performance improvement.

A belief often expressed by organizational leaders as well as consultants is that motivations and attitudes of those deciding on performance improvement is not that critical. After all, "something is better than nothing." Added to this is the well-accepted notion that organizations, teams, and people have a natural tendency to resist change and even the best performance improvement efforts will have limited success.[9] So why make a complicated and difficult process more so by requiring involved collaboration and diagnosis?

Misdirected efforts at performance improvement are often the cause of, or at least major contributors to, resistance to change. The use of Key Performance Indicators, discussed fully in Chapter 4, helps determine if the direction for improvement efforts is correct. When efforts are directed at the wrong issues, involve the wrong people, or do not have leadership support and participation, time and money are wasted, and frustration among participants can skyrocket. One sad version of this is the common attitude by members of the organization/team that the performance improvement effort is something to be endured, probably doomed to be ineffective. This attitude is even stronger if the current performance improvement program is simply the most recent in a string of failures. When this happens then the attitude can be, "something is often worse than nothing."

G. Suggested action steps for organizational/team leaders

Discuss the following topics with the top decision makers within your organization/team:

1. Do we have reasons for concern about performance in our organization/team?

2. Is the vision of the future of our organization clear to all involved in making it happen? Is it defined in a way that is specific and meaningful to us? Have we become satisfied with a two paragraph mission statement that does not clarify what we are about?
3. Do we have the expertise to clarify how well we are performing and to identify barriers to performance in reaching our goals. Do we have on staff people with expertise on performance improvement efforts? Who should be involved in deciding what areas, if any, need improvement?

End notes

1. Cummings, Thomas G. and Worley, Christopher G., *Organizational Development and Change*, 6th ed., South-Western Publishing, Cincinnati, OH, 1997, 3.
2. This definition is used by Achievement Associates, Inc. in a number of our performance improvement efforts. As is often true of intellectual capital, many organizations and performance experts have contributed to formation of the ideas.
3. Bassi, Laurie J. and Van Buren, Mark E., *State of the Industry Report*, ASTD, 1998, 26.
4. AMA/Deloitte & Touche, LLP, *Survey On Change Management*, 1994, 2.
5. Hall, Jay, *The Competence Connection: A Blue Print for Excellence*, Woodstead Press, 1993, xii.
6. AMA/Deloitte & Touche, LLP, *Survey On Change Management*, 3, 1994.
7. Robinson, Dana Gaines and Robinson, James C., *Performance Consulting: Moving Beyond Training*, Berrett-Koehler Publishers, San Francisco, 1996.
8. Schermerhorn, Hunt and Osborn, *Managing Organizational Behavior*, John Wiley & Sons, New York, 1991, 495.
9. Robbins, Steven P., *Organizational Behavior: Concepts * Controversies * Applications*, Prentice Hall, Englewood Cliffs, 1996, 723.

chapter two

Performance gaps and deficiencies: An overview

Chapter 1 stated that performance is the "definition and progressive achieve-ment of tangible, specific, measurable and personally meaningful goals," and then asked "goals to do what?" This had to be troublesome to readers concerned with moving straight through to understanding performance improvement. The reality is, however, that working to improve performance in any organization or team often involves the same back and forth, or iterative, steps. Do the leaders know what they want to accomplish, do they have a strategic vision? Sometimes yes, sometimes no. To clarify the issues here, let's put these concepts together in what would be an ideal way.

A. Model of ideal performance definitions by leadership

1. In an ideal world, leadership of the organization/team has clearly defined its strategic vision with all the elements described later in Chapter 7. They have answered questions such as: What business are we in? What do we do for whom? How can we be effective and efficient in achieving the vision we have laid out? In this ideal situation, the strategic vision is defined enough that strategic goals can be determined, clarified and communicated.

2. Standards, goals, productivity measures or other models of effective per-formance have been defined at the operational level.[1] These operational goals have been established as a result of the strategic goals. Everyone understands roles and responsibilities and who is responsible for what and when.

Unfortunately this ideal situation often does not exist.

At the leadership level, many organization and team leaders have only vague ideas about what they want their organization to be, now or in the future. In fact, many leaders are so wrapped up in solving daily crises and operational issues they have not thought through what they desire to achieve. Even if they have a plan or vision of their desired organization/team, they often do not communicate it clearly to others. Therefore, they live largely alone in their own vision and do not have a chance to gain support from those who could work with them to accomplish what needs to be done.

This frequent deficiency in leadership vision, either through lack of clarity and specificity or because of unwillingness or inability to communicate it to others, is not a condition reserved to any particular type of leader or organization. This sad and ineffective condition can be found in organizations that are for profit, not-for-profit, with leaders of all races and both genders and those from all age groups. One of the basic functions of management is to provide direction, but many leaders only provide marching orders. Sometimes even the orders are unclear.[2]

The deficiency in defining and communicating the strategic vision can originate from a lack of appreciation of the role of leaders in setting the long term direction of the organization. Sometimes, even when they appreciate the strategic role of leadership, leaders have the mistaken belief that performance will occur best when they keep their vision of the organization to themselves. The belief in this case is that the workers do not need to know the direction of their organization to do their work. In fact, some leaders believe that communicating the direction of the organization/team is a distraction and confusing to workers. There is a lack of understanding on the part of some leaders about how sharing organizational direction is essential in today's rapidly changing realities. In short, there is little appreciation of the potential motivational impact of what Drucker calls "policies to make the future."[3]

Even with the best intentions, however, some organization/team leaders have not figured out how to develop and communicate their mental picture of the organization as they want it to be, though they may want to. The following brief case will clarify the point:

Organization: Webster University

- Products and Services
  undergraduate and graduate level collegiate education
  related services such as books, residence facilities, etc.
- Markets
  graduate students usually employed full time
  undergraduate students with recent high school graduation
  a large number of foreign born students
- Structure
  main campus provides direction for distant sites
  numerous distance learning facilities within the U.S. and in a
    number of other countries as well
- Primary Issues/Problems
  new executive leadership at the central location of the university
    decided that each major unit needed a written plan for the future

the primary marketing unit of the university had recently been
   restructured to include some distinct functions previously
   operating independently from each other
common team direction and interaction was insufficiently
   established

The leader of the enrollment unit is a highly intelligent, educated and insightful person who only needed some help in clarifying the primary issues/prob-
lems and deciding how best to improve the situation. After a great deal of
consideration and diagnosis, she decided to engage in a process for developing a shared plan with 20 of her administrators and staff. The objective
was to develop a written plan that would be useful for building teamwork,
defining and evaluating performance, and providing rationale for future
budget requests.

The objective of developing a written plan was achieved. The challenge
then became the standard one, how do we get the strategic vision and related
goals into operation?[4] But the performance improvement initiative had succeeded in getting the leadership vision developed and shared by the top team.

B.   Performance at the operational level

The leaders' thorough and specific vision of the organization (in the above
case, a department or independent team) is the beginning basis for defining
performance, measuring it, and deciding if improvement is needed. The
leaders' vision should be clear enough to lead to specific goals which are
themselves measurable, and therefore partially define performance. But that
detailed and specific vision also provides the basis for performance definition
throughout the organization/team as the strategic vision and goals are
refined.

A number of times throughout the first few pages of this book I have
talked about leadership needing to provide a detailed, specific and thorough
vision which they use as the start of defining performance for themselves
and others in the organization. Who will be involved in defining this vision
will vary depending on the nature of the business, as well as the top person's
leadership style, personality, and attitudes. Some leaders want few others
involved, others want to include a large number in deciding on future
direction. The best guiding principle is this: "include those persons whose
support and acceptance is essential in achieving the vision."

A number of writers and consultants spend a good deal of effort in
discussing the relationships between participation in planning and specific
goal setting on the one hand, and the acceptance of the plan and goals,
commitment and motivation to achieving the plan and goals and job satisfaction on the other hand.[5] These are important questions because they
specifically deal with whether involving others in defining the strategic
vision and resulting goals and communicating this to others is worth the

effort. A few basic principles, supported by lengthy experience and substantial research, are warranted here.

1. Attitudes influence or modify behavior even though there is some debate on whether attitudes cause behavior.
2. Participation in planning and goal setting increases acceptance of those goals. (Attitude).
3. Challenging goals increases performance, as long as the goals are seen as achievable. Participation in goal setting, when done correctly, can lead to very challenging goals, which increases their acceptance, and therefore performance.
4. Some people are more motivated by participation in setting goals in which they are involved, others appear to be at least equally motivated by goals that are assigned. It is not easy to know in advance which group is which.
5. Specific goals are much more motivational than vague or general standards, e.g., "improve your performance."
6. Measurement and feedback to teams and individuals regarding goals for which they are responsible modifies behavior and probably increases performance over what it would be without measured feedback.
7. The above principles work about equally well with individuals or teams.
8. While job satisfaction does not produce performance (i.e., some satisfied employees are not performing at high levels), good performance usually increases job satisfaction.
9. There is a direct positive relationship between job satisfaction and retention of employees.

While these are important principles, one further point is critical. A leader, or the top leaders of a team or organization, needs the support and effort of those working for him or her to increase the possibility of improved performance. It is true that some employees will feel committed and motivated when they are involved in planning, or at least have the plan explained to them, while others may not. But the style of involving others in defining future direction influences acceptance and motivation as well, maybe as much as individual attitudes and characteristics. Involvement in defining the vision or goals of the organization/team can best produce acceptance or motivation if that involvement is seen by participants as real.

Once the detailed vision of the leader(s) is defined and communicated, it is time to set goals, establish standards, and identify models of effective performance for people throughout the organization or team. The process for this, and a lengthy discussion of goals, Key Performance Indicators, and standards is contained in Chapter 4. Here we will focus only on a brief definition and a few examples to clarify these important concepts.

1. A goal is an end result. There are many different types of goals: examples; profit margin, number of people to be hired, acceptable return rate on a product line sent to customers.
2. A standard is a very specific requirement, often in areas of production such as quality measures for a tangible item, or in areas of behavior required of employees in certain situations. For example, some companies using telephones for customer service support set a standard number of rings before the phone must be answered. Other standards in performance management might include a certain level of positive rating on teamwork using feedback from an employee's co-workers for measurement. Standards and goals are sometimes very similar.
3. A Key Performance Indicator is a measure of significant activity leading to an important goal. An engineering consulting company might use the number of project proposals submitted to key clients as a measure of activity towards a sales goal.
4. Models of Effective Performance are templates or descriptions of knowledge, skills, attitudes, and expected output required in an area of performance or attached to a particular position. For example, organizations may have detailed examples of management style they regard as exemplary and even required for their managers. The brief case study outlined below will clarify this.

Organization: A large U.S. retailer

- Products and Services
  department store products such as clothing, furniture, etc.
  support services such as credit
- Markets
  middle income families and individuals in suburban centers
  lower middle income families and individuals in some downtown
    stores
- Structure
  more than 200 stores organized into divisions essentially reflecting
    geographic location
  headquarters provided overall direction to the divisions
  each division had its own formal leadership
- Primary Issues/Problems
  customer service was not at the desired level

The organization had committed to quality customer service as an important advantage in the intensely competitive retailing industry. It had developed a set of standards or desired behaviors for sales associates in their interactions with customers. After rolling out the initial program, including training of the sales associates in good customer service, ratings from customers had improved. But then those ratings flattened out, and customer service was apparently no longer improving.

The customer service executives of the retailer and their performance consultants agreed on a process for improving customer service which focused on leadership within each of the 200 or so stores. Part of the education and training involved developing a model of effective performance, including specific behaviors and policies for leaders of the customer service process within each store. These managers were provided a model of effective customer service leadership, a model to emulate. Approximately six months later, the customer service scores average for all stores showed a very significant improvement.

C. Goals, standards, key performance indicators and models of effective performance: Measures of performance

Chapter 4 contains a lengthy discussion of goals setting, key performance indicators (KPIs), models of effective performance, and how they form the basis for measurement. KPIs are measures of activity that is important to achieving major goals. Here, the main point is that having these elements throughout the team or organization forms the basis for assessing current performance and whether performance improvement is needed.

A number of points needs to be made about a system of goals, KPIs, and models of effective performance.

1. The system provides a basis for knowing where performance ought to be based on the judgment of those setting the goals and establishing the models. KPIs help us know where goal attainment is and whether the activity to reach those goals is sufficient.
2. Knowing the current performance of the team or organization cannot occur without these elements, except at the very general or intuitive level. Having these system elements provides the basis for knowing current performance compared to desired performance. The difference is a gap or deficiency.
3. Without knowing exactly where performance is compared to where it ought to be, (gap or deficiency) there is little basis for knowing what to improve. This often leads to performance interventions based on fad or program popularity.
4. When you have a clear definition of performance and know the current status, then you have the basis for deciding what areas need improvement.

D. Identifying areas needing performance improvement

In the desirable situation, we have goals, standards, KPIs and models of effective performance in each organizational unit, department, team or subgroup. Performance in these areas are measurable either directly (e.g., are we opening new branches at the required rate?) or indirectly (e.g., is the store manager doing the things in our model that are necessary for building customer service?). This provides the basis for identifying significant gaps and deciding whether or not to engage in performance improvement.

The location of performance deficiency is commonly found in many areas. The following list does not include unclear or poorly communicated leadership vision for the organization which was discussed at the beginning of this chapter. Here we are assuming the leadership vision is in place. After the strategic vision is established and performance is reviewed, these following operational areas are frequently found to have performance gaps.

1. A specific department, a smaller unit or team within that department, where much of the rest of the organization is dependent on their performance. Example: sales/marketing teams, supervisors in manufacturing, etc.
2. The performance management systems, including: management does not use the systems, poor use of goal setting to start the performance cycle, or unclear connection between performance and promotion or financial rewards.
3. The leadership style is ineffective, or there is a lack of knowledge and skills in managing others, not only at the top level, but throughout the team or organization.
4. Work processes are poorly defined or ineffective in areas such as planning, hiring, or production.
5. The organizational structure retards production or customer service.

E. Conclusions

Deciding whether performance deficiencies or a gap exists that are significant enough to take action is not simple, even when goals and the other measures indicated above are well-established. Thoughtful judgment is required. Often organizational/team leaders are influenced by their status in the organization as well as their own motivations, personalities, and attitudes. A straightforward process for deciding what, if anything needs to be done and how to develop performance is critical to good decision making. Otherwise, performance improvement efforts can become diminished by organizational politics and egos. We turn now to discussion of a beneficial process for performance improvement decisions.

F. Suggested action steps for organizational/team leaders

1. Ask yourself the following question: Have I (we) defined our strategic vision clearly including: products and services we are offering and want to continue offering, markets/customers we serve, methods of marketing/selling what we do, etc.?

2. Ask the following question of others in your organization/team: Have we communicated what we are trying to achieve to you and others who are important to performance?

3. Review the following issues: Where do we have goals, standards, KPIs, and models of performance? Where are these elements absent? How severely is the omission of these performance measures hurting us? Do we really know how we are performing, in what areas the performance is acceptable, and where it is not?

End notes

1. The definition of models of effective performance is included in Chapter 3, and examples are contained throughout this book.
2. We discuss the differences and similarities between management and leadership in Chapter 9.
3. Drucker, Peter F., Management Challenges for the 21st Century. *Harper Business*, 1999, 73.
4. This was done through strategic planning. For a full discussion of the strategic planning process used here and in other cases identified in this book, see Chapter 7.
5. For example, Ramon Aldag discusses the relationship between job satisfaction and motivation to performance, while Smither talks about getting members of the organization to "reframe" the desired organizational situation before starting on an intervention. The latter author sees reframing as different from Organizational Development, an arguable point. See the list below for more examples.

* Smither, Robert D., *The Psychology of Work and Human Performance*, Longman, New York, 1998.
* Aldag, Ramon J.; Stearns, Timothy M. , *Management*, South-Western Publishing Co., Cincinnati, OH, 1991.
* Kotter, John P., *Leading Change*, Harvard Business School Press, Boston, 1996.
* Wall, Sobol, & Solum, *The Mission Driven Organization*, Prima Publishing, Rocklin, 1999.

chapter three

Deciding on performance improvement: Useful concepts and tools

If as we discussed in Chapters 1 and 2, the early motivations, attitudes, and decisions of the organizational or team leader are critical to success in performance improvement, then what can leaders do to increase the chances of success? The first rule is to keep an open mind and to recognize a critical fact: organizational and team leaders are often good sources of information about what is occurring, but they usually lack the knowledge, experience, and even objectivity necessary for deciding how to improve performance. A second look at an actual case discussed in Chapter 1 illustrates this point.

Company: Matheny Heating and Cooling Service, Inc.

- Products and Services
  HVAC equipment installation/repair and servicing
- Markets
  residential
  commercial
- Primary Issues
  declining and insufficient amount of business
  located in older neighborhood where the nearby markets were declining.

Recall from Chapter 1 that in original discussions with the two brothers who were owners and operators, it became clear that they believed customer service was the primary need. As discussions continued, it became clear that the owners wanted to improve the company's financial performance and brighten the long term future. But that had more to do with redefining their geographic market and re-thinking the mix of products and services than with customer service. While excellent customer service is always desirable, in this case providing improved customer service to a market declining because of population shifts was not the primary need. In fact, as an initial effort with this organization, it would have been a waste of time and money. The owners came to see the greater potential of strategic redefinitions of products and services and markets.

It is easy for organizational/team decision makers to start with the wrong motivations and go in the wrong direction for performance improvement. The tangible and intangible cost of those mistakes can be disastrous. Is there a process that can significantly increase the chances of proper motivation and direction for performance improvement?

A. Process for deciding about performance improvement: If / what / how?

Because decision makers often have limited understanding about how to improve performance in their teams or organizations, help is needed. Even if leaders have experience in developing performance, they benefit from having their perceptions and attitudes reviewed through discussions with others who are well intended and have knowledge about how to diagnose and improve performance. Thus, there is a valuable role for a performance improvement facilitator/consultant.

Any decision maker has the choice of deciding whether to bring in an outside performance improvement consultant or whether to depend on internal staff with expertise in assessing and developing performance. Using carefully chosen outside help means you may get someone with concepts, models, and experiences not possessed by anyone internally. The potential limitation, however, is that outsiders do not know the organization well. In addition, external consultants or process facilitators can be influenced in their judgment by the desire to "sell" more business and generate income.

Larger organizations, Anheuser-Busch, Inc., (ABI) for example, have internal performance improvement experts. This internal staff is frequently used for issues in the various plants and other units of the brewery. But outside consultants are also frequently called upon, usually because ABI leadership believes there is experience and knowledge they don't have internally.

Whether the decision maker decides to use inside or outside help in the process of deciding about performance improvement, the steps and skills required for success are largely the same. Surprising to many organizational leaders, while there are some differences in the challenges faced by outside versus inside consultants, the basic objectives in considering performance improvement are the same. Those basic objectives are as follows:

- Identification and definition of performance deficiencies and what, if anything, to do about them.
- Full collaboration and support between the leadership of the organization/team and the performance improvement facilitator (external or internal) to help make performance improvement programs work.

- Evaluation of the performance changes that occur when an improvement program is used and the development of systems and methods aimed at making the improvement.[1]

A current trend in overlapping areas such as performance improvement, training and development, and organizational development is the increased focus on doing things for business results.[2] The most common generic phrase to describe this is "performance consulting," which leads one to wonder what earlier consulting was about. The issue is to make sure that a process exists for clearly understanding performance goals, deficiencies, and opportunities for improvement.

Many models for deciding on and managing performance improvement exist, including management and training programs that focus on building team performance. A very clear and useful model is contained in a graduate-level organizational development textbook *Organizational Development and Change* by Thomas G. Cummings and Christopher G. Worley and is discussed below.[3]

THE GENERAL MODEL OF PLANNED CHANGE

Entering	Diagnosing	Planning/	Evaluating/
and	Models,	Implementing	Institutionalizing
Contracting	Data	Change	Change

B. The four stages

I.
The first stage of the General Model of Planned Change, *Entering and Contracting*, involves the initial contact between the organizational/team decision maker and the internal or external performance improvement consultant. The key requirement here is that a trusting collaborative relationship be built between the two sides in working on the issue at hand. What that means is that the organizational/team decision maker must be candid and thorough at dealing with questions about performance and providing information on deficiencies and issues, even when the realities are unpleasant.

Sometimes external consultants are desirable because internal performance improvement experts are seen as representatives of corporate leadership, and there is mistrust among employees about the motivations of that leadership. On the other hand, sometimes decision makers don't want outsiders to know about the organization's problems, so they are more open with performance facilitators from inside the company. Assuming knowledge and skills at performance improvement by both external and internal consultants, the most critical guideline in making the choice of a performance improvement facilitator is finding one who will generate the greatest openness in discussions.

The conversations in the *Entering and Contracting* stage are aimed at the following objectives:

- An initial broad definition of the performance deficiency and who and what factors are involved.
- A discussion regarding how to proceed with further inquiry into the issues of performance deficiency. Here tactical questions about methods of additional data collection are also answered.
- An introductory discussion of the role of the organizational/team leadership in supporting the diagnosis that follows this first phase.

The organizational/team leader should maintain candor in communication and recognize his/her critical role in the entire performance improvement process from beginning to end and beyond. The performance improvement facilitator/consultant has to establish and maintain commitment to using independent judgment in often sensitive discussions. That sets the stage for the consultant to add value throughout the entire process, no matter how far it goes. It is sometimes hard to remember that though the decision makers lead the team or organization that does not mean they completely understand the issues, causes or cures for performance improvement. The facilitator provides the best value by asking insightful questions, proposing alternative ideas and approaches, and disagreeing with the leadership where there is a basis for doing so.

The first stage of the General Model of Planned Change, Entering and Contracting, provides the foundation for the second stage, *Diagnosis.* If an effective collaborative relationship is built between organization/team leaders and performance improvement consultants, then Diagnosis has a good chance of working well.

II.

The second stage, *Diagnosis*, is both the most difficult and the most important of the stages in the General Model of Planned Change aimed at performance improvement. First, it is most important because if the diagnosis is not accurate, you may end up working on the wrong performance deficiencies and causes. Diagnosis is aimed at answering the following questions:

1. What are we trying to achieve with this organization/team? What is the strategic vision, what are the goals/objectives and standards resulting from that vision?
2. What area(s) of deficiency or needed improvement exist? We can't do everything at once, so what should our initial focus be?

3. What is the nature and what are the causes of performance gaps or deficiencies?
4. If we decide to get involved in Planning and Implementing Change, what intervention is most apt to improve the performance?

Words such as "deficiencies," "problems," "issues" often worry organizational decision-makers, particularly if they are in situations where the unit is apparently successful. Leaders sometimes seem to believe that their organization or team is perfect, or at least has little room for improvement. The question diagnosis answers is this: "Do we have room for significant improvement in performance? No matter how good we are now, can we do significantly better?"[4]

Sometimes organizational/team decision makers ignore deficiencies and problems for so long that their unit gets into serious performance trouble. This means that it is too late to make changes resulting in major performance improvement. Diagnosing performance on the basis of crisis can lead to premature judgments about what is really going on and rash judgments that "anything will make it better." The concept of "continuous improvement", made popular in quality programs, should be a guiding principle for organizational/team leadership.

Effective diagnosis requires three things:

- *A model(s) of effective performance* for the area of deficiency being reviewed is needed. What that means is the persons doing the diagnosis have to know (or be quick to develop models of) what organizational or team performance being scrutinized looks like when it is excellent. Different concepts and approaches provide examples of effective performance. What is the vision for this team at its best? What is the model of excellence? What are the benchmarks for best in class? What are the major standards for top performance? These are all questions describing the same thing: what does top performance in this area look like? Detailed examples of useful models of effective performance are presented below and throughout this book.
- *Measurement using data to compare the organization/team being reviewed to the model of effective performance* is the second requirement for making diagnosis work. Data can be obtained through interviews, surveys, observation or existing information collected by the unit during its normal operations. Which source to use is determined by practical considerations such as the area of concern and the size of the group or team involved. For example, if morale and turnover in a very large department is the concern, then a survey may be the most efficient data source because of the ability to cover a large group of people at a reasonable cost. If the leadership style of a small group of executives is the focus, then structured feedback from their direct reports and one-on-one interviews may be the best method of data collection because of the ability to probe deeper into individual style.

Many of today's organizations collect all kinds of data more or less connected to performance of teams and individuals. Data collection without some idea of what factors are crucial to effective performance, however, is often wasteful and random. Few significant changes occur. Too many organizations, for example, do surveys of employee morale or organizational culture, and then very little is improved. When this happens, frustration among those completing the surveys grows, and performance improvement efforts receive a negative reputation.

An example of an efficient survey based on a model of effective organizational performance will help clarify this point.

Organizational climate survey

1. Communication

"Identify the general degree to which communication is open, based on mutual trust and respect, clear, and effective in contributing to the achievement of organizational goals."

2. Responsibility

"Identify the extent to which members of the organization are given responsibility to achieve their part of the organization's goals and the degree to which they feel that they can make decisions and solve problems without checking with superiors each step of the way."

These first two questions of the survey are scored by having respondents place an A (for Actual) on a 9 point scale indicating where they think their organization/team actually is on the dimension. Respondents also place a D (for Desired) on the same scale for where they desire the organization/team to be. Group averages are calculated for both Actual and Desired, and the gap between the two averages is then reviewed and discussed. The key point is that these two questions reflect a model of effective performance, which includes at least the following principles:

"Organizational goals are central to the performance of the organization/team, and it is important to review communication and the responsibility given to members of the organization in the light of their contribution to the achievement of those goals."

To continue with this illustration, data collection about communication in an organization, one of the most popular survey topics, becomes more focused when we talk about communication pertaining to the achievement of goals. The same can be said for responsibility, or any other area of data collection.

Feedback to managers, sometimes called "360 degree feedback," has been increasing in recent years. But unless those collecting and reviewing the data have a clear model of effective managerial performance, the data is difficult to interpret, and usually the interpretation comes down to the attitudes of the person(s) doing the interpretation. That is the essence of what is meant by "subjectivity."

The concept of models of effective performance is based on the belief that we have principles and concepts that describe better or even best known performance for organizations and teams and within specific areas of responsibility (e.g., "good teamwork, and effective selling skills, or good supervisory skills"). It is also based on the belief that we can identify a model of effective performance for positions at the individual level, e.g., sales manager, CEO, etc.

This does not mean that there is "one right organization," a single most desirable organizational structure, or final answers about hierarchy in structure versus teams or matrix structures. The nature of the products and services, and the markets served, among other things such as size of the organization, all influence which structure will work better, and which ones will not work at all. But as Peter Drucker has noted, while there is no one best organization, "There are indeed some 'principles' of organization."[5]

Models of effective performance also do not mean that there is one right way to manage or lead people, with no consideration of their differences in personality, motivation, knowledge and skills. In fact, there is evidence, as you will see in Chapter 9, that the best model of effective managerial performance involves managing with consideration of the individual differences between those you manage. Similarly, selling has to be adjusted to the nature of the business. Selling pet food to thousands of veterinarians across the country requires prospecting while serving a customer who comes into the drug store does not. Still a guiding principle of effective selling in these and other situations is the ability to identify needs and wants and match products and services to those customer needs.

- Providing feedback to the organization/team decision makers comparing the current situation to the model of effective performance is the third element for productive diagnosis.

Obviously this is a critical decision point. The gap or deficiency between the existing situation and the model of effective performance may be small enough to leave things alone. If the gap is significant, some planning and intervention are required.

During diagnosis, the roles and responsibilities of the organizational/team decision maker and the performance improvement consultant are both critical and yet somewhat different. The consultant/facilitator keeps the process moving, contributes model(s) of effective performance, collects and presents the data and makes recommendations. The decision-maker provides data for the diagnosis, supports additional data collection as needed, and ultimately makes the decision about whether planning and intervention is to be done. This is a collaboration that is absolutely necessary for performance improvement to work.

III.

The third stage of the General Model of Planned Change, *Planning and Intervention*, only occurs if the diagnosis identified significant problems or deficiencies in the organization/team. In that case, the interventions identified in the diagnosis stage are extended and carried into action. This is not to say that additional interventions might not be added as the intervention is planned or even as it is carried out. The case outlined below will clarify the point.

Company: Motor Appliance Corporation

- Products and Services
    - customized electric motors
    - commercial and industrial battery chargers
- Markets
    - original equipment manufacturers such as golf cart manufacturers, etc.
    - users such as railroads, etc.
- Structure
    - corporate office in St. Louis
    - two plants, one in Missouri and one in Arkansas
- Primary Issues/Problems (identified during diagnosis)
    - managers need additional knowledge and skills about managing their employees and the production and sales processes

Discussions with the new CEO of this organization involved questions about strategic versus operational needs. Both performance improvement areas required development. The decision was to develop the top management team.

The management development program began, but some resistance to the program from selected managers existed from the beginning. It soon became apparent that there was an additional problem affecting performance which had not been uncovered during diagnosis. Specifically, there were major conflicts occurring between the sales force for one of the product lines and the production people located in a plant that manufactured that product line. Production believed sales was making promises to the customer that production could not keep. Production was charged with not working hard enough to meet deadlines and design quality specifications. The overall management development intervention was expanded to include work on inter-team communication and decision making.

A list of potential interventions for performance improvement will vary depending on who is providing the list. The interventions included below are focused on the human side of the organization or team rather than on physical equipment or information technology. The communication equipment and computer systems used by the organization/team, for example, will probably be part of a strategic plan, but the strategic intervention is still primarily about people choosing and then using those support systems. The following is a list of major potential interventions for performance improvement.

1. Strategic planning and cascading goals
2. Improving the hiring process
3. Restructuring to enhance process and/or process improvement
4. Improving effective teamwork for any team in the organization and building interteam cooperation
5. Developing intact teams in subject areas: management/leadership teams, sales teams, customer service teams, etc.
6. Performance management
7. Methods aimed at developing individual performance
8. Training, development and education[6]

IV.

Stage four of the General Model of Planned Change is *Evaluation*. Evaluation of changes in organizational, team, and individual performance is frequently given little attention in actual cases of intervention. As often as not this inattention is based on the belief that you cannot measure the soft skills such as how well managers manage or team members function. This problem with measurement starts with insufficient data collection during diagnosis and with not having a model of effective performance. If the diagnosis stage is insufficient in data or models of effective performance, then measurement of improvement is sparse.

Experts in training and development, most notably Donald Kirkpatrick, have been leaders in advancing evaluation of performance improvement. Many of the approaches and techniques useful in measuring performance improvement from training interventions, are also useful, with some adjustments, to evaluating performance improvement from other interventions. The key is measurement of specific outcomes. For example, have we accomplished the strategic objectives developed in our strategic plan? Have our efforts at improving the hiring process led to reduction in turnover or better performance by new hires compared to earlier hires?

Evaluation of the performance improvement effort is important for a number of reasons. First, you have to know what, if anything, in the intervention produced improvement so you can get started on finding ways to keep or "institutionalize" what has been effective. Secondly, motivation of those involved in the intervention for maintaining their efforts is more likely to occur when they are reinforced by real progress shown through evaluation.

Measurement of improved performance of individuals, teams, and organizations is not as "objective" as calculating profit at the end of the fiscal year. But if we know specifically what the organization or team is trying to accomplish, then measurement and evaluation of performance improvement efforts is somewhat more "objective." This means that it is possible to apply metrics to significant areas of performance, collect data, and usually get agreement on whether things have improved.

Donald Kirkpartrick has identified four levels of evaluation, though as mentioned above he describes them in their application to training programs.[7] The four levels he describes are:

- Evaluating reaction — how people felt about the intervention
- Evaluating learning
- Evaluating behavior
- Evaluating results

Part II of this book describes a number of performance improvement interventions in detail. This basic Kirkpatrick model with some adjustments applies to evaluating other types of intervention in addition to training and development.

Knowing what the organization or team should be trying to accomplish starts with a clearly defined strategic vision. In fact, I have said before, the strategic vision is the basis for defining performance and then deciding if improvement is needed.[8] If a department or team's efforts are almost entirely determined by the nature and direction of the overall organization, such as a production department in a factory, then the strategy of the broader organization should be the focus for defining performance for the smaller units. If, on the other hand, the team or smaller unit has a lot of freedom in deciding what products or services it will provide, internally or externally, then that team may well need its own strategy in support of the strategy of the overall organization. A marketing or selling team, for example, with many choices to make about what they are going to sell to which customer group, may benefit from their own strategy. Similarly, a large human resources department, with plenty of choices to be made on products and services to be offered to various groups inside of the organization, may also need a strategy for its own direction consistent with that of the corporation.

C. The organizational success model:[9] A model of effective performance

The model of organizational success presented in Figure 3.1 is frequently useful in helping organizations/teams decide what performance interventions, if any, they need to undertake. This model says organizations are comprised of two basic components: strategy and operations.

Organizational Success = Clear and Shared Strategy + Effective
and Efficient Operations

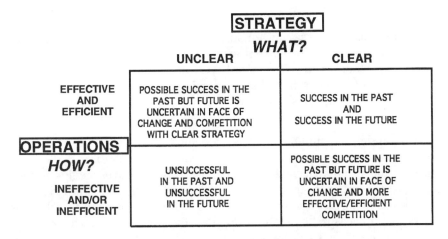

	STRATEGY	
	WHAT?	
	UNCLEAR	**CLEAR**
EFFECTIVE AND EFFICIENT	POSSIBLE SUCCESS IN THE PAST BUT FUTURE IS UNCERTAIN IN FACE OF CHANGE AND COMPETITION WITH CLEAR STRATEGY	SUCCESS IN THE PAST AND SUCCESS IN THE FUTURE
OPERATIONS **HOW?** **INEFFECTIVE AND/OR INEFFICIENT**	UNSUCCESSFUL IN THE PAST AND UNSUCCESSFUL IN THE FUTURE	POSSIBLE SUCCESS IN THE PAST BUT FUTURE IS UNCERTAIN IN FACE OF CHANGE AND MORE EFFECTIVE/EFFICIENT COMPETITION

Figure 3.1   Organizational Success Model.

The best use of this model is to begin with a review of the first principle, that of organizational success coming from clear/shared strategy and effective/efficient operations. Strategy is what products and services, marketing strategy and the like the organization/team is striving to obtain. Operations is how they manage, hire, provide customer service, reward and recognize, etc.

Diagnosis occurs using this model by discussing whether the strategy is clear and shared. A strategic concept statement, which is partly a general marketing plan, may be unclear for a number of reasons including new competitive activity, outdated products and services, shifts in the disposable income of the organization's market, or market decline.[10] McDonnell Douglas Corporation, for example, began to experience major strategic problems when the Cold War ended and the U.S. military reduced its defense budget. One of the major deficiencies in clarity of the strategic plan, however, is that it may exist only in the mind of the organizational/team leader and not be known by others in the organization who are needed to make it happen. Whatever the reason, if the strategy is not clear or shared by at least the top group in the organization or team, strategic planning is a possible intervention.

There are many designs for a strategic plan, some of them include a mission statement, and a list of values.[11] Our firm has found, however, that at least the following elements of a strategic plan are crucial to performance improvement.

- A Strategic Concept Statement: A detailed statement of what products and services, markets, methods of marketing/selling and distribution system, (among other things) the organizational/team leadership desires and intends to have in place at some point in the future. Again, this is usually referred to as the strategic vision and is thoroughly detailed.

- A detailed analysis of the current situation within the organization/team in its significant surroundings in light of where the organizational decision makers want to go (the Strengths, Weaknesses, Opportunities and Threats or SWOT analysis).
- A set of specific immediate strategic objectives for taking the first steps towards the strategic vision.

As a reminder, the strategic plan is a document that provides immediate and future direction for an organization, or, where appropriate, for a team. It sets up the insight that comes from diagnosis and improves the significance of measurement in both diagnosis and evaluation. The strategic plan includes strategic objectives and Key Performance Indicators and establishes the basis for measurable goals throughout the organization. The strategic objectives themselves can be made specific and measurable, so that there are evaluation of the progress of the top leaders of the organization or team. Some examples of specific measurement: develop a new product line for our fabrication division by June 30th; add thirty new branches within six months; or deliver sales training aimed at increasing business 20% by the end of the year.

We discuss strategic planning as an intervention process in Chapter 7. Here the primary point is that Strategic Planning at the top level of the organization, itself a performance improvement process, can set the basis for other performance improvement interventions because we begin to know what performance best serves the strategy. The strategy becomes the basis for immediate and future action, including, but not limited to, additional performance improvement interventions.

During the diagnosis stage, it may become obvious that the strategy is clear enough for the moment, and one or more operational areas needs performance improvement instead. Once that decision has been made, diagnosis may lead to the decision, for example, to improve the performance of the sales team. Here the data collection during diagnosis helps provide the basis for evaluation after the performance improvement intervention, as we look for improvement in the sales team's results. Discussion of a model of effective performance for selling is discussed in Chapters 8 and 9.

Whatever the performance improvement intervention, strategic planning or performance development of an operational area, measurement is important. The important phrase is "if you can't measure it, you can't manage it." Similarly, "if you can't measure it, you can't improve it." Clear strategy, Key Performance Indicators, and goals throughout the organization are the basis for measurement and performance management and, where needed, performance improvement.

D. Suggested action steps for organizational/team leaders

1. Decide who should be involved in discussing organizational performance.
   Who should be involved from your internal operations people?
   Who should be the performance improvement facilitator/consultant?
2. Decide how to keep candor in discussions at a maximum.
3. Decide how and when to start the initial discussion with the newly formed performance improvement group (entering and contacting). Structure these meetings enough where there is a brief agenda and notes kept to keep you on track.
4. Decide what diagnosis is needed. Are the issues macro or micro (the entire organization or a limited number of units?) What kind of data do you have; what kind of data needs to be collected?
5. What is the model(s) of effective performance here? What are the standards, goals, or performance objectives? If there are none, how to you get started developing them?
6. If there is a significant area of deficiency, what performance improvement intervention do you want to undertake? What are the costs and risks?
7. If you decide to proceed with the intervention, what type of evaluation should be conduct? How can you maximize chances to know how and where the intervention helped?

End notes

1. Part III of this book discusses stabilizing or "institutionalizing" performance improvement once it has been defined by evaluation.
2. Dubois, David D., *Competency-Based Performance Improvement: A Strategy for Organizational Change*, HRD Press, 1993. Robinson, Dana Gaines and Robinson, James C., *Performance Consulting*, Berrett-Koehler Publishers, San Francisco, 1996.
3. Cummings, Thomas G. and Worley, Christopher G., *Organizational Development and Change*, 6th ed., South-Western Publishing, Cincinnati, 32, 1997.
4. A problem is a barrier to a goal or objective. Therefore, diagnosis has to start with goals or directions of the organization/team and then get to the barriers or deficiencies. The concept of deficiencies in performance, sometimes called "gap analysis" is most clearly developed in diagnosis for training and development, but applies to other performance development needs as well. Discussion of this application to performance improvement generally is contained in Chapter 2. See: Blanchard, P. Nick, and Thacker, James W., *Effective Training*, Prentice Hall, Englewood Cliffs, 1999, Chapter 4; and two books by Dana Gaines Robinson and James C. Robinson, *Training for Impact*, Josey Bass Management Series, San Francisco, 1989, and *Performance Consulting*, Berrett-Koehler Publishers, San Francisco, 1996.
5. Drucker, Peter F. *Management Challenges for the 21st Century*, Harper Business, New York, 1999, 13.

6. For a useful distinction between training, development and education, see Dubois, David D., *Competency Based Performance Improvement*, HRD Press, Amherst, 1993, 4.

7. Donald Kirkpatrick, *Evaluating Training Programs*, Berrett - Koehler Publishers, San Francisco, 1998.

8. Kotter, John P., *Leading Change*, Harvard Business School Press, Boston, Chapters 2 and 5 in particular, 1996.

9. Credit for developing this model, along with a number of others in this book, goes primarily to W. C. "Mike" Weaver, President of Achievement Associates, Inc.; Achievement Associates, Inc. also gives credit to Benjamin B. Tregoe and John W. Zimmerman, *Top Management Strategy*, Simon and Schuster, New York, 1980.

10. Drucker, Peter F., *Management Challenges for the 21st Century*, Harper Business, New York, 1999, 58.

11. Bob Wall, Mark R. Sobol, and Robert S. Solum, *The Mission Driven Organization*, Prima Publishing, Rocklin, see Chapter 3, 1999.

chapter four

Performance improvement and goal setting: Making the strategic vision happen

While a clear and shared strategic vision from the leadership of the organization/team is the foundation for performance improvement, it is only the beginning. When the vision regarding products and services to be offered, markets served, methods of marketing and the like are clarified, we have begun the trip toward performance improvement. But many thorough strategic visions have stalled because people further down in the organization do not understand the strategy, are unclear about their responsibility, or don't buy into that strategy. Having goals and Key Performance Indicators throughout the organization is essential for driving performance improvement toward achieving the strategic vision.

A. Characteristics of effective goals

Much has been written in general management literature about goals in organizations and teams. The following characteristics of effective goals is based on practical experience as well as a good deal of literature and is presented in the context of performance improvement throughout the organization.

1. Goals should be tied to or "cascade from" the detailed strategic vision developed and communicated by the leadership. The discussion of strategic planning in Chapter 7 discusses this further.
2. Goals should be clear and *specific*.[1]
3. Goals need to be *measurable*, thereby providing the basis for evaluating performance at any point, including progress or lack of progress in performance improvement.
4. Goals should be *achievable*, but challenging.[2] The common phrases capturing this characteristic are "stretch goals," and "out of reach but not out of sight." This characteristic is the basis for enhancing performance from where it would be without these stretch goals.

5. Goals should be *realistic*, meaning that they should not ask the team or individual with the goal responsibility to do the impossible or to do something inconsistent with the nature and business of the organization.
6. Goals should be *timed*; they must be attached to deadlines for their accomplishment.

The catch phrase used to communicate these principles is SMART goals (the word corresponding to each letter of SMART is italicized in the above list.) Surprisingly, many organizations make little use of performance goals or objectives, and when they do they are often not smart. This is true despite the fact that it is impossible to know with certainty the status of performance unless goals and Key Performance Indicators are set and used to measure and manage performance. Sometimes organizations, even Fortune 500 companies, provide strategic goals or objectives from the top, but they are so general, they are difficult to define and measure. They become less useful to teams and individuals below the top level in identifying what they should be doing to help the organization achieve its strategic vision and goals.

B. Sources of resistance to goal setting

I have found three primary reasons why organizations, teams and individuals resist setting goals. The first, and perhaps most influential, is fear of failure. The logic is, "if we are not pinned down to specific goals or objectives, we cannot be held accountable and punished if we do not perform." At the leadership level of the organization/team, leaders can be fearful of knowing how well things are going, perhaps simply to avoid feeling bad.

Experience has shown that when employees avoid goal setting, it is not usually motivated by laziness, but rather by lack of confidence, or fear that leadership is a punishing group looking for reasons to hassle employees. For people finding themselves in this situation, there is a better way to deal with it other than avoidance of goals. Why not work to develop specific goals for yourself or your team that meet the above characteristics of smart goals? Then, if you perform, it is much harder for unkind leadership to harrass you. If you do not, at least you can figure out how to improve. For organizational/team leaders who avoid goals so as not to feel bad about failure, setting and reaching challenging goals can provide increased self confidence and satisfaction. Otherwise, doubt and insecurity continues.

The second reason why organizational/team leaders do not use effective goal setting is lack of appreciation of the benefits from using such a system. The benefits are numerous, and have impact beginning with the top level of the organization, to the functioning of teams, and finally to individual performance. The list of benefits is provided below.

A third reason for resisting goal setting is the belief that in today's rapidly changing world setting goals is impossible because of the need to be adaptive. The reasoning goes something like this: "How can we set a target if the demands of the daily situation are hard to predict? New business opportunities may occur, or customers may suddenly have new demands, or some technical breakdown may happen." The problem is that unless the organization or team has goals, they become slaves to activity through reacting to changing circumstances. The better approach is to set goals at the team and individual level, but include methods and targets for being responsive to crises and opportunities.

C. Benefits from organizational/team goal setting

Organizations, teams, and people tend to do things because they see desirable benefits. This is especially true of initiating goal setting which requires new behavior and attitudes. The following are benefits of a goal system that are well-supported by research and experience.

1. Goals provide focused action for achieving the organization or team's strategy. Goals provide direction and focus for performance.
2. When teams and individuals have effective goals they are clear about what is expected of them.[3]
3. Goals are motivational, helping to elevate performance because most people like the feelings of success that come with reaching challenging and yet realistic goals.
4. Goal setting and achievement of challenging goals builds job satisfaction and self confidence from those performing well.
5. Setting goals can make work more fun and can help relieve boredom because the goals become like the score in a game with the fun coming from meeting the challenge.
6. Goals provide the basis for measuring and evaluating current performance and for evaluating progress or lack of progress with performance improvement efforts.

D. Cascading goals

As mentioned earlier, goals that cascade from the top of the organization/team down to smaller units and individuals are helpful for driving performance. A case study will illustrate the point.

Company: Tone's Brothers

- Products and Services
    processed spices
- Markets
    retail chains
    food manufacturers
    cafeteria management companies

- Structure
    plant of approximately 1000 people
    about 200 managers, sales and customer service personnel
    unionized factory workers
- Primary Issues
    the need to connect corporate strategy to specific goals in production

This company wanted to reduce inventory, increase the number of inventory turns, and more closely connect product runs to the marketing strategy. To achieve this, I conducted a hands-on seminar with the production managers and supervisors using the following steps.

1. We discussed personality and behavior in a team setting by providing each participant with a development focused personality profile and used that information to discuss individual teamwork strengths and challenges. On their own, each participant identified things they would do to increase their team functioning as the group set its production goals.
2. We then reviewed the corporate strategic goals and systematically identified goals at the production level that would help them do their part in supporting corporate strategy.
3. We identified barriers or problems faced by production in achieving their department goals and developed suggested action plans for overcoming those barriers.

The team of production department supervisors left the sessions with a list of specific goals and actions to work on. As always, how well these were achieved had a good deal to do with department leadership support and persistence. In general, this particular company made a good effort at following through in achieving production goals that were set.

This case illustrates a useful process for cascading goals. First, leadership of the organization/team needs to make their strategic vision and resulting goals clear. Following that, each group or team within the larger unit should go through the process of identifying their goals and objectives and activities necessary for them to play their part in the achievement of the overall strategy.

E. Special case: Large teams in big organizations

We have indicated that sometimes teams have the independence to consider developing their own strategic vision. When that occurs, they then provide a strategic vision to sub-teams within their larger team, and the cascading of goals process continues as usual. A brief case can make this clearer.

Organization: Pet food supplied to veterinarians and breeders

- Products and Services
    - dog food
    - cat food
    - specialized pet foods, litter
- Markets
    - pet superstores
    - regional chain stores
- Primary Issues
    - the need to clarify their goals and actions in support of Ralston's corporate strategy
    - the need to build the team towards increasing cohesion and performance

This team recognized that it had significant independence in making strategic decisions. Specifically, the team could decide which products offered by its company it wanted to sell to whom, which specific customers should be pursued, what was the best sales and marketing approach, and how to distribute the product once it was sold. While the team had to develop this strategy consistent with the strategic direction and policies of the company, setting its own direction within that broader context was a key to higher performance.

The team began with strategic planning, developing a written plan involving a strategic vision of the desirable and intended future of their team, assessing its current situation, and setting immediate goals for working on the strategy. What followed, however, was particularly creative. This top team had additional meetings to discuss what each manager would do to carry his or her share of the strategic objectives. In addition, they identified and discussed what they and their direct reports needed to do to carry on their daily activities as part of the larger company. In short, each member of this top team committed him or herself and his or her direct reports to both strategic goals and daily operational goals.

F. Cascading goals down to the individual

In the large team case discussed above, the planning process was completed when the team members developed and communicated individual action plans for achieving both the strategic and operational goals. Other individuals within their sub-teams also developed individual action plans and operational goals. Thus the connection of the levels was complete from corporate goals to large team goals, through the goals of the smaller teams, and finally to the individual level.

Many performance experts have recognized the influence of goal setting on team performance.[4] It is perhaps less well understood that goals, when developed correctly, also can have a positive impact on individual employee performance. Goal analysis is often the recommended technique for competencies in areas of individual attitudes, but in general, organizational/team leaders are much quicker to set goals for teams than for individuals. Many performance management systems(appraisals), for example, in all kinds of organizations lack clear goal setting at the initial planning stage when the individual employee and his or her manager plan work during the performance cycle. This is discussed in detail in Chapter 10.

Setting goals at the individual level is sometimes resisted because it requires specific persons take responsibility for making things happen. At times, goals are effectively "cascaded" from the top level to the team and/or sub-team level but are not integrated into individual action plans. This can be an oversight on the part of those responsible for moving the goals through the various levels of the organization, or it can be resistance from individual employees. But the benefits of goal setting set forth above are no less true for the individual employee than they are for the total organization or the teams that comprise it.

Resistance to taking responsibility for goals on the part of individuals may be caused by a desire to avoid accountability, as we said earlier. But as often as not, it is caused by the goals simply being "forced" on employees, rather than discussed and agreed to in a collaborative manner. Leaders sometimes fear collaborative goal setting because they believe employees will "sand bag" by setting goals too low. Actually, our experience and varying research projects have indicated the reverse, employees are more apt to set goals too high, making them difficult to reach.

As the Ralston Purina case above illustrates, there are many different types or categories of goals. There should always be financial goals, even for not-for-profit organizations such as government agencies because they still have budgets. In fact, the most commonly found goals in organizations are their budgets and the expense and income targets. But other goals should exist as well such as production goals, selling goals, hiring goals, goals for creating new products and services, customer service goals, and human resources development goals.

G. Measuring performance: Goals and key performance indicators

Just as the strategic vision is the starting point for performance improvement and setting goals, the strategic and operational goals that flow from strategy are the starting point for measurement of performance. A goal is a desired end result. A major goal may have a set of shorter term or smaller goals leading to the desired end result. The shorter term goals still need to be smart. For example, if the end or major goal is developing and rolling out a new product by a certain date, then one of the shorter term goals might be hiring a qualified research and development engineer.

There are always a series of important activities required to reach goals, short term or longer term. If the activities are not undertaken successfully, the goal will not be accomplished. Measurement of success in important activities leading to goals is done through Key Performance Indicators (KPIs). KPIs are a limited number of performance measures that answer the following questions: are we doing enough of the right things to reach our goals? If not, what is the corrective action to be taken?

While measurement of performance at the organizational, team, and individual levels is necessary for performance evaluation and performance improvement, no organization or team need spend huge amounts of time or resources on collecting data for measurement of goals or KPIs. Only a few key performance indicators need to be chosen for measurement. For example, if our goal is a certain level of sales volume, then the KPI may be new leads. If our goal is a certain level of employee morale measured by a survey, then our KPI may be employee turnover.

A key concern is picking the right factors to measure, not measuring less important factors and not measuring too much. Picking the right factor to measure is important for two reasons. You have to measure factors where trends in the right direction on the KPI is indicative of progress on the goal they support. In addition, when organizations or teams use a KPI to measure activity, they are reinforcing it in the minds of the employees. Usually, organizations and teams tend to get more of what they are measuring.

Establishing long term and shorter term goals permits measurement of performance through assessing goal achievement. Establishing KPIs, when we correctly identify the factors to be measured, also permits measurement of performance leading to goal achievement. Thinking back to the General Model of Planned Change in Chapter 3, it should be clear that performance measurement is important, though in varying degrees, in all four stages of the model.

During the first stage of the model, *Entering and Contracting,* organization and team decision makers are going to be clearest about the need to get involved in deciding on performance improvement if they have a number of significant measures of current performance. Discussions about how to proceed with diagnosis, and what areas of the organization should be reviewed, can be more focused by early performance measures. In diagnosis, additional data may well need to be collected, but existing data on performance is a good place to start during these initial discussions.

As we have said, the second stage, *Diagnosis,* is the most important and the most difficult step in deciding on and delivering efforts at performance improvement. When we come to understand what areas of performance need review, and what models of effective performance apply, then the need for additional data may become apparent. There are four potential sources of data in any organization or team: surveys, observation of performance, interviews, and data already existing in the organization (intrinsic data).

Decisions about which source(s) to use is a purely practical matter driven by decisions in areas of concern, what data already exists, costs, and existence of standards or models of effective performance.

H. A brief case study: Another model of effective performance

Company: An engineering fabrication and process consulting company

- Products and Services
  fabrication of process control instruments, analyzer systems
  consulting in mechanical, electrical, and chemical processes
- Markets
  manufacturing companies
- Structure:
  about 170 employees
  a small number of divisions, one focused on manufacturing, another on consulting
- Primary Issues/Problems
  dramatic growth through doubling their size in the past 5 years
  recently purchased by a huge company with resulting problems of adjusting strategy and procedures

After early conversations with this CEO/owner focusing on the above issues, we reviewed the Organizational Success Model (see Chapter 2) to decide whether his primary issues were strategic or operational. We concluded that the issues were principally operational, with employee morale and relationships between offices and divisions being central. We then talked about the following Model of Effective Performance.

Effective Performance = Clear/Shared Goals + Good Attitudes + Knowledge/Skills + Motivation[5]

The interviews with leadership provided data indicating that the primary issues were attitudes. The specifics of the attitudes and morale held by the 170 or so employees, however, needed further clarification and definition before diagnosis could be completed and action decided upon. We developed a preliminary intervention which included collection of survey data from the employees regarding challenges, opportunities, and morale. The top teams from all divisions and offices then would discuss what to do about significant attitude problems.

As this brief case shows, data collection and measurement is often useful during *Planning and Intervention*, the third stage of the General Model of Planned Change. During the intervention stage, the effort should be to directly tie measurement to performance improvement by using the data to trigger improvement actions. So, using the above case, the engineering company's top team had the opportunity to use data on morale and attitudes in their efforts at improving these factors and enhancing communication and collaboration between divisions and groups of employees. In this case, bad morale and distraction by take over issues was believed to have negative impact on performance.

The final stage of the General Model of Planned Change is *Evaluation and Institutionalization* of desired changes. This, of course, only occurs if there is a performance improvement intervention. Clear goals and Key Performance Indicators provide the basis for measurement which itself helps us decide what in performance has been improved and therefore what we want to keep or "institutionalize."

Part III of this book discusses institutionalizing performance improvement, that is making it an on-going part of the organization. Since we are focusing in this chapter on goals and measurement, however, it is important to make one major point very clear. Specific goals and related KPI measurements are essential for making performance improvement interventions last.[6] A brief case will make this concept clearer.

Company: Scottsdale Securities, Inc.

- Products and Services
  online discount broker
  broker assisted trades as needed
- Structure
  approximately 100 corporate personnel
  approximately 100 branches spread throughout the U.S.
- Primary Issues/Problems
  lacked shared strategic vision
  CEO who was the founder and primary owner had at least a partial vision of what he wanted to accomplish, but it was not fully developed and not completely understood and/or accepted by his corporate and branch managers.
- Actions
  conducted a thorough strategic planning process with the twenty or so top managers at the corporate level.

The leadership of this organization agreed to engage in strategic planning. The early discussions with this leadership group indicated that the diagnosis of the need for strategic direction was correct; there was a lack of shared strategic vision among the leadership group. The CEO and his top advisors understood that agreement on a strategy, along with specific goals or objectives to accomplish that strategy, was essential to continued growth and success in the firm. The CEO could not do it by himself. Nineteen strategic objectives requiring immediate action were established at the end of the strategic planning process. Individual and team responsibility for these objectives was established primarily through the top managers volunteering to champion strategic goals/objectives they felt best able to accomplish.

Strategic goals or objectives are basically oriented towards long term development of the organization. For this discount brokerage firm, daily operational issues, such as customer requests or technical problems for their business which is dependent on automation, often led to periodic delays in progress on their strategic objectives. But the leadership of the organization continued to review and evaluate progress on each of the objectives. Within a year, all but one of the nineteen strategic goals was either completed or resulted in major progress towards achievement of it.

I. The role of feedback

Establishing cascading goals and Key Performance Indicators to accomplish the organization's strategy sets up the possibility and requirement for feedback to those responsible for accomplishing the goals. Feedback to teams and individuals is both troublesome to those worried about not achieving the goals and yet essential to evaluating current performance and identifying any need for performance improvement.

Feedback to a team or department about how it is accomplishing its goals should be done in a way in which all members of that team know how their group is doing. A quote from teamwork expert Tony Montebello succinctly states the point:

> Feedback has powerful motivational and developmental value for teams. It lets the team and the members know how they are doing so they know what to do more of, less of, or differently. Feedback can be about performance results, variances from plan, coordination with other units, customer satisfaction, or effective and ineffective behavior on the team. Depending on the type of feedback required and desired, there obviously are different sources such as internal customers, management, suppliers, and the team members themselves. But for feedback to be effective, it must be timely, specific, balanced and candid.[7]

How much more effective will the feedback be when it is tied to specific team goals and Key Performance Indicators? The same is true for performance at the individual level. Another performance expert, Robert Smither, notes:

> One fact that researchers have demonstrated repeatedly over the years is that goal setting improves performance. ... Goals act as motivators, and research in both applied and experimental settings shows that difficult goals produce higher levels of performance than either no goals or simply instructions to "do your best". ... When workers know a particular level of performance is expected of them, they are motivated to try to reach that level, even though it may be difficult. In addition, when group members are committed to a goal, performance of the group improves.[7]

As this quotation from Montebello indicates, how feedback is provided is directly connected to how it is accepted or resisted and therefore how it affects performance. Our suggested criteria for providing feedback to teams and individuals about their performance is as follows:

1. Feedback should be specific and where possible tied to specific goals or Key Performance Indicators. Numerical data and specific examples are helpful.
2. Feedback should be sincere reflecting honest attitudes and opinions from the provider of the feedback. Few things are as frustrating for employees as having a manager who just read a new book or attended a management seminar and starts to provide ineffective feedback he or she doesn't really believe.
3. Making feedback timely is also critical. Receiving feedback on an occurrence some weeks past reduces the impact.
4. Feedback should offer suggestions for performance improvement where there are deficiencies. This is often a problem for managers providing feedback because they may not have an understanding of how performance can be improved. Obviously, referring the employee needing help with performance improvement to someone with experience in developing people is one possibility. In addition, sometimes setting specific goals for what performance improvement actions should be taken will lead both the manager and the person receiving feedback to find ways to make improvements.

Feedback is more apt to be taken seriously and serve as a trigger to action when the team or individual are committed to the goal or measured activities being discussed. Earlier we argued for using a collaborative goal setting process to enhance commitment to the goals and motivation for achieving them. While there is some evidence that employees can have as much commitment to assigned goals as to ones they help set, their participation in goal setting usually has no negative consequences and often helps with acceptance or motivation. For example, it is well-established that commitment to a goal works best when the goal is consistent with the employee's values and when he or she believes success in goal achievement is possible. Active collaboration between the employees and manager while setting goals allows for the opportunity to discover if this goal meets these requirements while simply assigning the goal may not.

A brief case illustrating an effective and collaborative goal setting process followed by feedback illustrates a number of the points discussed in this chapter.

Organization: A privately owned real estate company managed by the owner

- Products and Services
    - brokered real estate for buyers and sellers
    - purchased land and built apartments and other buildings for investment and sale
- Markets
    - home owners selling or buying real estate
- Structure
    - multiple offices in small towns in a primarily rural area
    - corporate staff including owner/manager and financial executive who managed budgets and conducted research on interest rates and local economic trends.
- Primary Issues/Problems
    - owner/operator wanted to make sure that all branches of his geographically separated organization were motivated for achievement
    - he also wanted to have a strong indication of probable financial performance for his total organization based on input from each office and real estate agent

This CEO/owner recognized the importance of gaining commitment and motivation to goals. The process he used for achieving this was to meet with each individual real estate agent and their local office manager to set sales goals for each agent. The CEO had information developed by his financial vice-president pertaining to projections on interest rates, which dramatically affect real estate sales, and other occurrences in the local economy such as openings of new large businesses, etc. The goal setting discussions were collaborative, and as often as not the CEO and local office manager recommended reducing the sales goals suggested by the employee agent. They did not want to encourage frustration by letting these aggressive agents set goals that were both out of reach and out of sight.

The goals formed the basis for feedback to individual agents. The combination of goals for the agents in a specific location, formed the basis for the goals for the local office, and the total of all the office goals combined to form corporate goals. This permitted the CEO to make financial decisions about buying and building real estate for his company investment.

It should be remembered that commitment to a goal is not the same thing as motivation to achieve it. But motivation to achieve a goal will probably not be very strong if the goal is not accepted as desirable. So commitment is necessary for excellent performance but is not sufficient. What else is needed? In general, the employee must come to believe that working hard to achieve an organizational/team goal for which they have responsibility will lead to their receiving rewards and recognition they value. More about this in Chapter 9 and the discussion of leadership and its role in performance improvement.

J. Conclusion

Performance requires at least three elements: smart goals that employees are committed to and are motivated to achieve, key performance measures on progress in goals achievement, and the resources to achieve the goals. These resources include tangible factors like equipment, financial support, as well as effective leadership which helps develop and clarify goals. But it also includes having the right people in the right jobs, and helping the people achieve the needed knowledge and skills for success in doing the work. We turn now to discussion of these elements and how best to conduct performance improvement actions when there are significant performance gaps.

K. Suggested action steps for organizational/team leaders

1. Discuss the following question with your top advisors. Do we have clearly established strategic goals, and do we communicate them well to those further down in the organization?
2. Consider this question: What Key Performance Indicators do we have? Which ones do we need?
3. Does our performance management system have goals and KPIs included?
4. Answer this question through input from employees: What strengths and problems exist in the ways our managers provide feedback?

End notes

1. Szilagyi, Jr., Andrew D., *Management and Performance*, Scott Foresman and Co., Glenview, 1981, 134.
2. Aldag and Stearns, *Management*, 2nd ed., College Division, South-Western Publishing, Cincinnati, 1991, 428.
3. Aldag and Stearns, *Management*, 2nd ed., College Division, South-Western Publishing, Cincinnati, 1991, 428.

4. For example, see: Cummings, Thomas G., and Worley, Christopher G., *Organizational Development and Change*, South-Western Publishing, Cincinnati, 1997, 99. Montebello, Tony, *Work Teams That Work*, Best Sellers Publishing, Minneapolis, 1994, 322, and Aldag and Stearns, *Management*, College Division, South-Western Publishing, Cincinnati, 1991, 428.
5. This model is used by Achievement Associates, Inc. in many individual and team diagnostic situations.
6. Cummings, Thomas G. and Worley, Christopher G., *Organizational Development and Change*, South-Western Publishing, Cincinnati, 1997.
7. Montebello, Anthony R., *Work Teams that Work*, Best Sellers Publishing, Minneapolis, 1994, 322.
8. Smither, Robert D., *The Psychology of Work and Human Performance*, Longman, New York, 3rd ed., 1998, 225–26.

part two

---

*Performance improvement:
Taking action*

chapter five

Building a learning organization

In Part I of this book, we discussed a number of key points which are recapped here.

- Organizational leaders feel a powerful need for performance improvement in the United States. This felt need is driven by a number of factors ranging from intense and growing competition, to the changing nature of work, to the importance of building shareholder value in corporations, to the desire to be competent and successful.
- Decision makers from all types of organizations/teams initiate performance improvement efforts with motivations than can start the efforts in the wrong direction, missing opportunities for sufficient diagnosis of the situation.
- Performance starts with the strategic vision of the leader(s) of the organization/team. That vision, which includes what products/services provided and for whom, is often poorly detailed or not shared with others in the group or organization who are needed to make it happen. Clarifying and detailing a thorough strategic vision is achieved through open discussions with those knowing the organization and/or how to improve performance.
- Once the leader(s)' strategic vision and the accompanying strategic goals are communicated to the organization/team, the opportunity exists for identifying performance requirements at the operational level. That is done through goals, standards, Key Performance Indicators and models of effective performance. These operational measures provide the basis for knowing the current status of performance.
- There is a process (General Model of Planned Change) for making decisions about performance improvement and managing improvement efforts from diagnosis though intervention, evaluation and institutionalization of improvements. The diagnosis stage of the process seeks to identify areas of performance gaps between desired goals and standards and actual achievement.

- Performance requirements are most effective when they are cascaded as goals or standards from the strategic vision of the leader(s). Efforts at performance improvement also need the same leadership support and specific goals and other measures to evaluate progress and make it permanent.

When used correctly all of these key points allow the organizational/team decision maker to answer the following questions.

1. What is good performance in my group or organization?
2. In what areas is our performance good enough to be acceptable, at least for now?
3. In what areas do we have performance deficiencies or gaps?

The eminent Peter Drucker recently wrote a brief passage which captures the essence of Part I.

> Every organization operates on a theory of Business, that is, a set of assumptions as to what its business is, what its objectives are, how it defines results, who its customers are, what the customers values and pay for. Strategy converts this Theory of the Business into performance.[1]

But strategy and achievement of the strategic vision is achieved by people throughout the organization/team performing at work in ways that accomplish goals, meet standards, and are consistent with models of effective performance in areas of importance. So people have to know how to do their work. Since strategy is dynamic, and there is constant and accelerating change in the environments of organizations, people have to keep up with the performance requirements in their organizations and teams. Ultimately the need for performance improvement by all employees is endless.

A. The learning organization: A starting point for performance improvement

Many hard driving organizational/team leaders, motivated by success and money, politically astute, and somewhat cynical may start this chapter with a good deal of skepticism. "Learning organizations" sounds academic or at least extravagant. What does this really have to do with performance improvement anyway?

A vast number of experts in management, performance improvement and related fields have been arguing for at least the past decade that we live in an information age, where what we know is the primary key to how well we as individuals, teams, or organizations succeed. As Peter Senge notes,

"The organization that will truly excel in the future will be the organization that discovers how to tap people's commitment and capacity to learn at all levels of the organization." Senge also uses a quote from the head of planning at Royal Dutch Shell to dramatize the importance of learning, "The ability to learn faster than your competitors may be the only sustainable competitive advantage."[2]

We started Chapter 1 by discussing the powerful felt need for performance improvement in our society which has led to efforts at improving organizational, team, and individual performance in all kinds of organizations. There are a number of interventions available for improving performance: strategy, restructuring, performance management systems, or training and development, for example. But they all require learning on the part of those involved in these interventions. And the rapid change in technology, the way employees do their work, and the requirement of being responsive to clients or customers, means that employees at all levels, whether involved in formal performance improvement programs or not, have to continuously learn as part of doing their jobs well. This is at least part of what Drucker and others mean by "knowledge workers."

The following summarizes some of the main points that come both from research and experience at building learning organizations.

1. Workers at all levels and in all kinds of organizations have to know a great deal about things ranging from how to operate computers, to best methods for dealing with customers, to the most recent improvements in technology for performing the jobs they do. What they need to know changes constantly.
2. The lean organizations so common today reduce the influence of supervision and management for workers at all levels, increasing the need for those workers to be even more productive than before, doing the work of more than one person.
3. A lean organizations also means that many workers must develop self management knowledge and skills.
4. Workers at all levels have to be prepared to change jobs, add significantly to the jobs they do, even change careers. Learning is a life long process.
5. The role of managers and supervisors is undergoing change in which they are required even more than before to mentor or teach those under their direction is order to enhance performance. Increasingly, workers solve problems and managers and supervisors help them learn how and find the resources for the problem solving.
6. Organizations who are leaders in building learning cultures usually want their employees to have more than knowledge and skill in the core competencies related to their current jobs. They also want understanding of the culture, values, and traditions of the organization, as well as knowledge of competitors and best practices at least within the area of expertise of the employee.[3]

7. Some organizations have come to recognize that learning in areas that are somewhat personal and may affect life outside of work is important. For example, education on personal finances is increasingly common and courses on goal setting and time management are everywhere and may be applied to personal areas of life as well as work.

8. Organizations benefit by providing many learning opportunities for their "knowledge workers," partly to improve productivity, and partly to keep them from going elsewhere.

9. Learning activities need to be connected to a clear understanding of what the organization needs to accomplish (its strategy) and what the employee knows how to do, does not know, and needs to know now and in the future.

10. The points made above are important to work organizations of all types, not just large for-profit corporations.

To summarize the above, a high performing organization must have employees, essentially from top to bottom of the structure, who have the following characteristics: They recognize the need to continuously learn and develop their knowledge and skills; they know they will be asked to add to what they do, take over someone else's job temporarily or as a permanent move, and teach others how to do things; they are willing and motivated to go beyond the core requirements of performing their current job to activities ranging from coaching and counseling, to helping design the strategy of the organization/team. The organization must both hire people who have these characteristics and provide them with the learning opportunities to make the above a reality.

B. The model for growth: Learning and skill development

In performance diagnosis and improvement, models play a critical role in helping us understand what is occurring now, what causes the current performance level, and how we can make it better. The Model for Growth in Figure 5.1 helps with understanding what areas of individual performance need emphasis. In short, it helps us to be specific about learning and skill development and methods to be used for that learning and skill improvement. The elements of the model in Figure 5.1 are as follows. The first is **Identity**, which is who you are as a person. This includes personality, motivation, attitudes, values, needs, etc. Identity, combined with **Knowledge and Skills**, drives behavior.

There are three sources to **Identity** in this model. The first source is *genetic inheritance* or heredity: what we all get from our parents, grandparents and others up the family tree who have influenced our basic personality and other factors. Psychologists have long debated how much of who we are is inherited from our family lineage, and how much is learned after birth, the "nature vs. nurture" issue. No one seems completely sure about all the things that are genetically caused, but some things clearly are: gender, some aspects

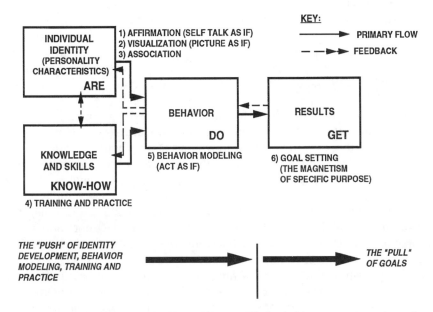

Figure 5.1   Model for Growth. (From Weaver, W. C., Achievement Associates, Inc., St. Louis, © 2000. With permission.)

of physical type or dimensions, predisposition towards certain illness, and possible some personality traits such as assertiveness.

The second source of identity is *body chemistry*. Our bodies are chemical factories, both producers of chemicals such as endorphins and users of chemicals in things such as metabolism. Chemicals influence mood, such as the natural high that comes from endorphins and the chemicals that are thought to produce some types of depression. While as human beings we have certain basic chemical processes, we can, unlike our genetic inheritance, influence our internal chemistry by what we consume and by exercise.

The third source of identity is *conditioning*, the attitudes, behavioral habits, values and motivations we have learned as we have dealt with our environment and responded to it. Attitudes are "habits of thought," conditioned ways of looking at our environment and elements in it. Attitudes also predispose us to act or behave in certain ways, what are called behavioral tendencies.

A number of learning programs and activities are focused at least partially on the identity of the learner. This is a brief sample of some of the better known learning tools:

1. Personality assessments aimed at hiring and or individual development. Examples include Myers Briggs and the DISC (See Chapter 6 on hiring, and Chapter 10 on improving performance at the individual level.)

2. Feedback instruments which evaluate the person's motivation, values such as honesty and attitudes towards co-workers or the organization. Most morale surveys rely heavily on assessing elements of identity.
3. Feedback instruments on leadership or supervisory style which look at management philosophy and attitudes towards workers, e.g., theory X vs. Y, authoritarian or laissez faire style.

The second element of the Model for Growth is **Knowledge and Skills**, what we know (concepts and information) and what we can do (skills). Knowledge and skills together are called "know how," and it takes both for performance to be effective. If we understand something, but cannot translate that knowledge into how we behave, it cannot help our performance. Conversely, if we can do something, but do not really understand it, we have trouble transferring that skill to different applications or situations. For example, if a person succeeds at dealing with a difficult customer, but does not understand what worked, he or she will not have much success in transferring that successful performance to other customer situations.

Many (even perhaps most) learning programs and activities are focused on the employees' knowledge and skills. Training and development programs for managers, customer service providers and sales people, for example, teach concepts and techniques for performing in those roles. When we teach employees new computer systems, or conduct on the job mentoring and training, our goal is to build knowledge and skills.

Translating knowledge, concepts, ideas, and information is one of the great challenges in training and development. Very often, things learned in the classroom are poorly transferred to the work situation, if at all. More about the specifics of that issue in Chapters 11 and 12. Here, the key point is that people learn to do things differently and more productively (skills) through practice. Remember that human identity is formed partly through conditioning whereby we develop attitudes, habits, values and motivations through dealing with our environment. Developing new skills is in large part working to eliminate habits that do not work and replacing them with more desirable and productive habits. This is the methodological basis for the wise phrase, "Repetition is the mother of learning."

The first two elements of the model for growth, **Identity** and **Knowledge and Skills**, produce the second element, which is our **Behavior**, or what we do. So, who we are, and what we know how to do, results in our behavior, which is what we do. Behavior is learned habits, application of knowledge and skills in doing something, the essence of our performance. It is the basic reality of how we function as human beings, team members, or members of an organization.

Many learning activities and programs focus on actual behavior. For example, individual performance assessments done as a basis for individual learning plans, are usually based at least partly on how the employee actually works. Feedback from one's supervisor during performance reviews should be based on actual performance or behavior, rather than vague references to "traits." Teaching people how to provide superior customer service, sell more effectively, or manage others well, should have a focus on suggestions about effective behavior in these roles. Leadership style, to carry this one step further, is the behavior of a manager in relationship to his/her direct reports.

Who we are (**Identity**), what we know how to do (**Knowledge and Skills**), results in what we do (**Behavior**), which leads to the fourth element of the model for growth, what we get (**Results**). As long as we keep being who we are, knowing only what we now know how to do, we will keep doing what we have been doing, and therefore usually keep getting the same results. Building a learning organization is establishing an environment, programs and activities, and on the job opportunities for individuals to learn in ways that help them better understand, improve and make use of their Identity, their Knowledge and Skills, their Behavior, and their Results.

There are a number of basic techniques available for people to learn and grow within the framework of the Model for Growth.

1. Techniques for learning and growth related to identity

When it comes to our identity, feedback from valid personality learning instruments, attitude or value profiles can provide us with useful understanding of who we are. But what action can we take to change our identity if desirable? Is it possible to change ourselves? The answer is yes, but with great effort and determination. Remember that identity is at least partly attitudes and habits developed through life. Those attitudes and habits are reinforced in our identity by the way we converse with ourselves (self-talk) and with others. This is called affirmation.

We talk to ourselves constantly. The way we do that is a reflection of our basic attitudes and values, and at the same time, a reinforcement of those elements. If an employee is worried about performing a particular job, he or she will often reinforce that concern by worrying about that deficiency. If an employee is confident about his or her ability to do something, or learn to do it, then that tendency is reinforced as well. An example will help clarify the practicalities of this.

Our firm worked with a sales person who changed careers fairly late in life and moved from being technically focused to increased emphasis on dealing with and selling the products and services to customers. We helped this individual through sales training, and also pointed out that attitude was a key to successful performance. One day this person made his first sales call on his own without any assistance. When he entered the office of the CEO, he found himself "affirming in the negative," talking to himself about his lack of experience in selling and relatively poor understanding of the products/services he was representing. But remembering the importance of attitude, he caught himself in poor affirmation and began to think through the positive opportunity he was in. With some struggles and a good deal of effort, this fledgling sales person made a substantial sale to the prospect. Had he continued the negative attitude with which he started, failure was a high probability.

Identity is also affected by how we see things, our "vision" of something. When there are discussions of the vision of organizational leadership, it is a discussion of the mental pictures the leader has about what is and what he/she wants the organization to be. However, all people have visions. The sales person discussed in the case above had a vision of failure and had to change it to have any chance of success.

Additionally, identity is affected by association, or who we spend time with. The qualities of our team or department co-workers, as well as the people we spend time with outside of work, affects our self image, how we feel about ourselves. People in customer service, for example, who have to spend the majority of their work day with angry customers often become hostile and frustrated themselves.

The lessons to learning organizations for helping people improve their use of their identity is clear: help them better understand who they are through feedback. Then teach them skills for beginning to affirm in a positive and productive way through repetition of messages, feedback about unproductive and productive attitudes, and providing performance feedback to them that is sincere, specific, and timely.

2. Techniques for learning and growth in knowledge and skills

Conceptually, the techniques for growth in knowledge and skills are straightforward; learn and then practice what you have learned. From a teaching perspective, the process is show them or help them learn, provide practical opportunities for doing what was learned, then provide feedback on success and problems. This of course means that the effort is to get the knowledge and skills into the behavior of the learner. From the learner perspective, the key is to find any and all opportunities to apply what has been learned and make efforts to improve how it works. Unfortunately, many learning experiences do not have structured practice situations.

3. Techniques for learning and growth in behavior

It is possible to work directly on a learner's behavior by teaching him or her better ways to do things. When a journeyman welder shows an apprentice techniques of the trade, that welder is "modeling" skills to the learner. When an organization develops its supervisors by teaching what effective supervisors do, they are providing a model to be emulated. We can learn to behave more effectively in some role by identifying a desirable model and learning to emulate what they do. The use of this approach is widespread in management/leadership development, customer service/selling development, technical education and training, and even efforts at improving the way a team functions.

4. Techniques for learning and growth in results

It should be noted that working on Identity, Knowledge and Skills, or Behavior, is done to get better Results. They are the targets of learning and development, because we believe that improving understanding and action in those areas produces better results. But there is also a basic approach for changing results directly. It is goal setting.

Remember that goals are an end result: specific, measurable, achievable, realistic, and timed. When we set or are assigned goals to which we are committed, we begin to work at figuring out what behavior is required to achieve that goal. Sometimes the behavior required is learning new knowledge and skills to make the goal happen. If through learning new behavior and knowledge and skills we accomplish the goal, then our confidence about our ability in that area, is strengthened. Therefore, the goal achievement has helped with Results, Behavior, Knowledge and Skills, and ultimately, an attitude of confidence that is part of our Identity.

A real life example will help illustrate the point. A manager decides that she or he wants to improve teamwork in his or her department by having problem solving meetings on a regular basis. The manager discusses the idea with team members and decides to meet Friday afternoon from 2 to 4 p.m. To meet this activity goal, the manager has to change her or his behavior on Friday afternoon by establishing, attending, and participating in the problem solving activity. As a result of the meetings goal, the manager also gains new knowledge about how to conduct problem solving team meetings, what is going on in the department, how the team members feel, and so forth. Knowledge and skills have improved.

If, to continue this example one step further, the meetings are conducted effectively, the manager may like the results in terms of morale and specific problem solving actions. It is possible that the results will be good enough that it becomes a part of the manager's attitudes about leading. The manager's self-image and other attitudes about what works in management, even his or her "vision" of effective leadership, may well have undergone change. He or she has learned to be a better manager. His or her identity has grown.

The model for growth focuses on learning by the individual. Sometimes learning occurs in a group, team building with an intact team for example. The model for growth still applies as a way of understanding what areas receive focus for learning in teams. Teams are made up of individuals working together for common goals and purposes. Ultimately, it is the individual team members who must buy into the common goals and learn what it takes to achieve the team goals.

C. Building learning organizations: *Two case studies*

Two case studies will help illustrate how organizations can implement many of the concepts, tools, and techniques identified in this chapter. As always in this book, these are actual cases where I have had direct involvement.

First case

Organization: Landshire, Inc.

- Products and Services
    packaged sandwiches, burritos, etc.
- Markets
    convenience stores and gas stations
    miscellaneous quick food retailers, bars, etc.
    provide product to outlets in 12 states
- Structure
    centralized preparation and packaging of their product
    corporate headquarters near the plant
    depots located strategically throughout their 12-state geographic
    market for short-term storage of products
    route drivers and supervisors deliver the product to retail outlets
    who sell to the end users
- Primary Issues/Problems: The owner and the CEO together had two
  primary elements in their visions of the organization. They wanted
  to encourage the development and growth of their employees, and
  they wanted to maintain organizational growth and financial success
  in a high stress route business.

The owner leaves operations largely to the CEO and other top managers, but
has strong feelings about treating employees well, as does the CEO. They also
are very successful as a business, adding outlets and even entire states in an
on-going manner. But they knew that route businesses had characteristics that
could produce intense problems. Their production and delivery schedule is
unforgiving; every day outlets demand deliveries. Failure to deliver can result
in the convenience store chain shifting to other food suppliers.

In addition to schedule challenges, route businesses, particularly ones
such as Landshire with a large geographic area to cover, face problems that
come from the physical separation of their employees. There are route drivers
operating in numerous locations in each of these states, and local supervisors
attempting to carry out corporate policy. Coordination and communication
is a special problem. The schedule and the geographic spread of Landshire's
customers results in production, selling, delivery and service needed at
hundreds of locations across 12 states on a six-day-a-week basis. There is
little time to spare.

The leadership of the organization knew they had some performance
problems related to attitude and motivation. Interpersonal conflicts in the
organization sometimes took an unreasonable amount of time and attention
away from production and delivery. Turnover, always high in route businesses because of the stress, was probably higher than it needed to be. Other
operational issues also existed, despite the fact that great strategic position-
ing within a high volume growth industry (convenience stores) made them
financially successful. The leadership decided to undertake a series of performance improvement activities.

- They were involved in a small number of thorough surveys to define the specifics of their operational issues as people within the organization saw them. They wanted to learn more about where they were as an organization, so that performance improvement could be targeted and progress could be measured.
- One of the most dramatic results of the surveys was that there were leadership style issues, some lack of clarity about directions and policy, and the existence of pockets of fear of punishment if things did not go well. Continuing these conditions was unacceptable to the two top leaders.
- As a result, Landshire has engaged in a series of learning and performance improvement interventions aimed at improving leadership style throughout the organization and providing clearer direction whenever possible. These have included the following:

  1. Involving their managers at all levels through learning effective management and leadership styles, tools, and techniques. This has included providing a number of on-site management development programs that require participation.
  2. Evaluating whether employees are using the management and leadership knowledge and skills that have been provided. This has been done fairly regularly during performance feedback sessions from managers to their direct reports, themselves managers of others.
  3. Using a collaborative approach to establishing selling, servicing and financial goals within each of the geographic areas comprising the total market. Feedback on achievement of these goals within each area is then regularly provided, and actions taken as needed, recognition and reward is freely given when deserved.

The executives at Landshire probably would not think of themselves as a "learning organization" in many of the ways listed earlier in this chapter. But they are headed in the right direction for a number of reasons.

1. The top leadership of Landshire is committed to using learning programs and developing knowledge and skills as a primary way of creating the kind of organization they envision.
2. They recognize that learning, from management, to selling, to technical operations, must be directly related to their strategy and the nature of their business. They recognize and have dealt with the need to provide learning opportunities to employees spread across 12 states and performing a wide range of functions.
3. They have sought to tailor learning to the knowledge, skills, identity issues and strengths specific to each team or individual involved in the learning experience.
4. They have shown incredible persistence in providing learning opportunities to their employees, some 8 years after their initial surveys identifying culture issues. Similarly, they have been determined to

reinforce learning through setting goals and providing feedback on how well people use what has been learned.

As of this writing, Landshire continues as a successful business. And they continue to work on training and developing their people in seminars, classrooms, and on the job.

They have the most essential characteristic of a "Learning Organization", they know that learning is a life long process, and they persist in finding ways to provide their people with opportunities for learning related to performance improvement.

Second case

Landshire, Inc. has accepted the challenge of developing the performance of their widely dispersed people through learning opportunities. By comparison to the next organization, they have relatively few resources to expend on learning. Anheuser-Busch, Inc. is at the other end of the resource continuum, spending millions of dollars per year to provide learning programs and activities.

Organization: Anheuser-Busch Companies Inc. (ABI)

- Products and Services
    many brands of beer, adventure parks, packaging
- Markets
    sells beer in more than 80 countries
    commands 47.6 percent market share in the U.S.
- Structure
    24,000 employees producing, marketing, and delivering product
    operates 14 breweries, 12 in the U.S., headquarters in St. Louis
    field sales force located in regions across the United States

Primary Issues/Problems: A-B had developed a group of training and development programs and activities that were spread throughout the organization. Each of these efforts was largely developed independent of the others. The objective was to pull these units together into a "corporate learning function" to accomplish the following goals:

1. Make sure the programs and activities were tied directly to corporate strategy.
2. Increase the reach and penetration of the collection of programs through coordination and sharing of efforts.
3. Increase the "relevancy," tie the programs and activities directly to position related "core competencies" and other supportive learning opportunities.

The decision makers at Anheuser-Busch created an organization reflecting a new learning alliance between some pre-existing programs. The learning alliance, entitled the Anheuser-Busch University (ABU), has the following units:

College of Production and Engineering	College of Sales/Marketing	College of Leadership and Professional Development
Technical training for the breweries, information technology, corporate engineering, purchasing and all SAP related training	Developmental experiences for sales/marketing, brand management, media groups, field sales, and distributors	Provides leadership and professional development company-wide. Develops core curriculum. Provides executive education

ABU has a number of features that makes it a model for learning programs in larger organizations.

1. Their method of organization serves the overall strategy of the corporation, a major emphasis on production quality followed by extensive sales and marketing. The leadership and professional area supports the emphasis on production and marketing, and works across the company to ensure consistency of message.

2. The learning activities that are conducted are based largely on assessment of what individuals need to enhance their long-term performance. A map of the performance strengths of the units in which employees work is conducted by the leader of the organization, and the College of Sales and Marketing (CSM) is conducted by the leader of that organization and the CSM. This "strategic map" focuses on the group's strategic plan, their structure, teamwork and team building, leadership strengths and issues, and review of processes. In addition, each individual to be served by the CSM conducts a self-assessment on required knowledge and skills for their position, and how they stand in mastering that information and those abilities.

3. The CSM in particular provides learning opportunities for important organizations in its distribution chain, the wholesalers and retailers receiving their products. This characteristic of providing learning to groups outside of the employee population is a distinguishing characteristic of "Corporate Quality Universities."[4]

D. Summary and conclusion

The two cases used here are about organizations with very different resources. That was done to make a point. If the leaders of organization/teams believe that learning and developing their people is crucial to performance success, they can find ways to do it. A-B University has hundreds of people involved in conducting its programs. They have an educational satellite, and they make extensive use of computer-based training, and interactive television.

Landshire conducts on the job training for production and sales/marketing employees, and they find external programs and resources for much of their learning effort.

The two organizations, dramatically different in size, both share a common characteristic. They recognize the value of learning in performance improvement, and they make a determined effort to maximize learning for the benefit of their organization and their employees.

E. Suggested action steps for organizational/team leaders

Answer these questions.

1. What do we know about ourselves as people? What are our attitudes, motivations, and basic personality characteristics? How can we come to better understand who we are and how to get more out of our identity?
2. What areas of work do we have sufficient knowledge and skills? In what areas do we need improvement in knowledge/skills?
3. What areas of behavior are most significant to our organization/team: managing, marketing/selling, customer service, technical support, financial, etc.? In which of these areas do we have sufficient positive models to move our behavior in the correct directions? In which of these areas do we need to find new models of effective performance? How do we find them and then install them in our organization/team?
4. Are our goals clearly established at the organizational/team level and the individual level? How do we increase goal clarity? Do we need to re-do our performance management system to increase the role of goals and KPIs?

End notes

1. Drucker, Peter F., *Management Challenges for the 21st Century*, Harper Business, New York, 1999, 43.
2. Senge, Peter M., *The Fifth Discipline*, Doubleday Currency, New York, 1990, 4.
3. Meister, Jeanne C., *Corporate Quality Universities*, Richard D. Irwin, Inc. New York, 1994.
4. Meister, Jeanne C., *Corporate Quality Universities*, Richard D. Irwin, Inc. New York, 1994, 33.

chapter six

Selection and performance: Process, tools, and techniques

A. Introduction and overview

Most organizational/team decisions makers understand, at least intuitively, that selection of people for positions and how they perform are closely linked. After all, performance has the best chance to be excellent if the organization has the right people in the right job, and it will suffer if they do not.

But this intuitive wisdom about the importance of good hiring is sometimes clouded by the following issues:

1. Organizations often have trouble knowing what personality characteristics, behavioral traits, or competencies are required to do well in a specific position.
2. Even when they know the above, which is infrequent, those involved in hiring for organizations often have trouble matching the applicants' characteristics to what is known about job requirements.
3. When a good applicant is found, the match with job requirements is made a hire occurs, and everyone is happy. Often the happiness lasts for only a short period of time. As a CEO business associate once was heard to say: "I hired him and then his ugly brother showed up." "Ugly" referred to performance.
4. Despite having experience with and performance information about current employees, many organizations are poorly motivated in transferring and promoting existing employees. These motivational tendencies include: placing a problem employee somewhere else, in hopes they will do less damage or straighten out; taking a person performing well in operations, and making them a manager because they have operational skills (e.g., sales person to sales manager,

production person to supervisor); and depending solely on a person's performance record in one job even though they are being considered for a very different position.

Despite these frustrations and issues, using an effective selection process remains critical to organizational and team performance for a number reasons.

- First, the objective of selection/hiring is to find a person who "fits" the requirements of the position. This means the person should have the motivation, personality, attitudes, and values that make performing well on the job possible. If any of these identity factors are inappropriate for the position, performance will suffer. Toward the end of this chapter, we will discuss "normative profiling," an approach based on profiling identity elements that are related to job performance and then developing a process for measuring those elements in job applicants.
- Secondly, managers are usually looking for applicants with knowledge and skills necessary for quick ability to perform in the position. Some organizations believe that the identity factors above are the most critical considerations in hiring, and when they find the right hire they can teach the knowledge and skills necessary to the position very quickly. That approach is most appropriate for hiring for less complex or entry level jobs. In general, the organization will want to hire people who already have knowledge, skills and requisite experience when they fill positions such as marketing/sales managers, top sales people, top managers, and technical or production employees.
- Third, organizations and teams are made up of interdependent workers. How well any individual performs is at least partly dependent on how the individual interacts with and supports co-workers.[1] For example, the supervisor on the first stage of the production line may need to collaborate with the supervisor of the second stage to solve some processing problem. Hiring needs to pay attention to how the applicant will fit with the culture and personalities of the people with whom they will work.
- Hiring is very expensive. Experience and research shows that out of pocket costs for hiring a manager averages about one year's annual salary. And if hiring is expensive, poor hiring is even more expensive. No one is quite sure how to precisely measure the costs of a poor hire when it leads to the person leaving later, but the calculation of costs should include re-hiring expenses plus lost production during down time.

If effective hiring, making sure that the applicant fits the job, is important for the organization/team, it is also critical to the applicant. How many people have had their careers derailed, spent months, years, or decades unhappy at work because they were in the wrong position? One author notes that studies show most workers would not choose the same line of work again.[2] All in all, it makes sense for those hiring and those being considered for a position to work together to decide if the fit is good for both.

B. Sources of information for hiring decisions

Because of the difficulty in knowing exactly what is needed for excellence on the job, because people are complex and often difficult to assess as applicants, and because personal life influences job performance, hiring is more like gambling than a science. The key is to improve the odds at making the right decision in the hiring gamble, which is principally achieved by gathering as much information as possible about the applicant. The following are sources of information important in hiring decisions.

1. The place to start is in identifying or gathering information on the identity characteristics (personality, motivation, values and attitudes), knowledge and skills, behavioral style, and output goals required for excellent performance in the position being filled. This is more than the traditional job description, which is often outdated because jobs change quickly, or it was written by one person, usually the supervisor, with no confirmation of accuracy. Developing a thorough understanding of the individual characteristics and competencies required by the job takes a systematic approach.[3]

The Model for Growth (Figure 5.1), discussed in Chapter 5, is useful for purposes of clarifying job requirements categories and is presented again as Figure 6.1. Information on job requirements for each of the boxes in the Model for Growth is important. Can anyone doubt that the identity factors of personality, motivation and basic values play a part in performance? While the importance of knowledge and skills will vary with the complexity of the position, skills at dealing with people and working in teams is almost always important. In addition, a profile of behavior or style required for success in the position is important. Finally, some positions, production, project management and selling as examples, require a strong commitment to output goals and deadlines. There are methods available for gathering information on all of these areas of job requirements.

Once the requirements of the job have been clarified, information about the applicant can be used to see how they fit with those requirements.

2. After review of the information on resumes and application forms, often done proforma, checking references should usually be done next. There are many challenges with reference checking as most Human Resources people know. Those challenges include the fear of lawsuits which leads them to be cautious above providing information. The basic recommendation is to use questions that are largely factual and focus on the knowledge, skills and

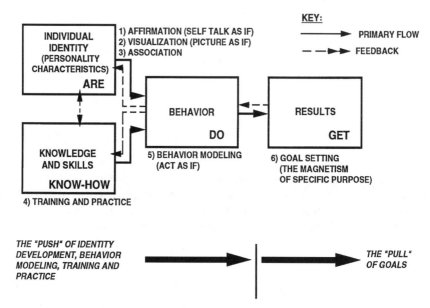

Figure 6.1 Model for Growth. (From Weaver, W. C., Achievement Associates, Inc., St. Louis, © 2000. With permission.)

goals of the applicant. So, in asking for references from a former employer, the following questions should provide some information. "Does Mr. X have a good understanding of marketing research techniques?" "Does Ms. X usually meet deadlines in submitting reports?" The one factual question that can be asked about overall performance which is frequently answered is: "Would your organization re-hire Ms. X?"

Doing reference checks is time consuming, but the information can be valuable. Knowing how someone has performed in the past is a guide to possible performance in the future with the new organization. But it should always be remembered that previous performance of an applicant was conditioned by at least three factors: how he or she was managed, whether how the job fit, and any personal life factors that may have influenced performance on the job.

3. Interviewing usually follows somewhere after the reference checks have been completed and the applicant seems to be a reasonable possibility for hiring. While reference checks can be difficult to conduct, interviewing has even more traps. The ineffectiveness of interviewing usually comes from one of two basic realities: most managers are not good at interviewing, and some interviewees are very good at "reading" what they think will play well with the interviewers and then providing it rather than saying what they really think.

A few basic interviewing guidelines can help hiring managers improve the results:

- The interviewer should talk less, particularly early in the interview. Get the applicant to talk more about him or herself before he or she can begin to figure out what the interviewer is looking for. The interviewer is trying to gain insight about who the applicant is and what he or she has done. Later talk more about the organization and what you want.
- Use many "open ended" questions, especially early in the interview. This also helps the manager avoid legal traps set up by an over-use of specific questions.

> Example: "What do you believe are characteristics of effective customer service in a retailing environment like ours?"

- Pursue information and follow open ended questions by going at least three layers deep in the topic.

> Example: "You said one characteristic of effective customer service is good credit policy. What do you mean by good credit policy?"

> Example: "In general, how would you provide customers information on the new credit policy you have just outlined?"

- Plan the interview and make sure you have reviewed all reference checks, resumes, and application data before the meeting.
- Asking questions on background, legal ones of course, provides some information on identity: motivations, basic values and attitudes, and the like.
- Asking questions about recent employment experiences tells more about knowledge and skills, work behavior and success at achieving work goals. When pursuing these areas, remember to use open ended questions followed by specific probes to get the details.
- Ask the applicant what information he or she wants about your organization. This is usually best done toward the end of the interview. Remember, a good fit between applicant and position is to everyone's benefit.[4]

Because of the ability to see the applicant face to face and pursue information with good asking and listening, interviewing is an extremely valuable source of data in hiring. But even when hiring managers follow guidelines for effective interviewing, like those above, applicants may "role play" away from who they are towards what they think you are looking for. Back to a slightly modified version of the quote from a CEO business associate used earlier in this chapter: "We conducted an effective interview, I hired him, and his ugly brother showed up."

4. Personality profiling, attitude surveys, and various types of knowledge and skills tests are additional important sources of information for hiring. The use of an instrument to measure one or the other of these areas is widespread today after decades of decline in use, probably because of fear of law suits.[5] The types of instruments available is best understood in terms of what the seek to measure.

Knowledge and skills tests, sometimes called ability tests, are in abundance for more technical areas. One supplier catalogue lists tests in the following areas: Office related tests, computer skill and aptitude tests, and industrial and mechanical tests. For example, an actual test within these categories is the SRA Test of Mechanical Concepts, which purports to measure basic mechanical ability and knowledge of common mechanical tools and devices.[6]

Attitude surveys and interest tests, which are similar in what they measure, are also used a good deal by organizations when hiring. The objective here is to measure a significant component of identity, attitudes and beliefs, which are related to behavior and performance. We discussed the general relationship between attitudes and behavior earlier; here the issue is to figure out which attitudes are connected which performance in the organization and/or position.

One useful attitude survey is the Pass III, the Personnel Assessment Selection System, a tool that was originally developed in response to the outlawing of polygraphs. This tool, which is completed in about fifteen minutes by the job applicant, measures attitudes in three areas thought to be directly related to performance. Those areas are alienation (attitudes towards work, supervisors, employers and co-workers), trust (whether the applicant believes in the value of trustworthiness), and drug and drinking related attitudes. An actual example of the use of this tool will help illustrate its benefits.

Organization: Site oil

- Products and Services
  convenience food and gas marts
- Markets
  the Midwest
  Florida
  automobile drivers
- Structure
  wide geographic spread with its outlet stores

    managed by a small corporate staff
    active human resources function
  • Primary Issues/Problems
    concern with retention of store personnel and reduction in theft
    of their store products

The human resources director decided to try the Pass III along with another potential tool. Being inquisitive and thorough, he sought to correlate both tools to retention of employees. He found the Pass III to be useful in hiring people who were more apt to stay with the company, at least partly because they had fewer "alienated attitudes" about being managed and working for others.

Personality profiling is potentially the most beneficial to effective hiring because it measures characteristics in the basic identity of the person. At the same time, it is one of the most controversial of the "testing and profiling" approaches. There is a lot of misunderstanding regarding what personality profiles do and cannot do.

A concern about personality profiling, still existing but less strong than a decade ago, is the issue of "legality." In general, a valid profile used consistently, with no attempt to illegally discriminate, will not lead to legal damages for organizational users. Courts have recently decided that if an organization shows that its hiring process does not have "adverse impact on protected groups," then the hiring process will be assumed to be non-discriminatory. "Adverse impact" is the ratio between applicants and hires for protected and non-protected groups, and the ratio for protected groups should be at least four-fifths of the ratio for non-protected groups to be legal. Of course, anybody can be sued for anything in today's litigious world. The key requirement is to be ethical and non-discriminatory in hiring so that legal actions are less likely to occur or lead to adverse decisions to the organization if they do occur.[7]

Using a valid personality profile consistently for any hiring occurrence can provide useful information on the personality, motivation and other aspects of the applicant's identity. The concept of validity is somewhat confusing, but essentially there are two important types of validity. First, there is *construct validity*, which essentially is determining that the profile measures what it purports to measure, and that there is a reasonable correlation between what it measures and how the person completing the profile behaves. There are various ways to validate a profile, including observation of behavior or correlation of results with another valid instrument that measures the same behavior. The goal here is to determine if they act as expected based on their profile results.

The second type of validity is *concurrent validity*. This has to do with the correlation between an applicant's results on the profile and performance in a particular job. There are different levels of concurrent validity. The most general level is when a profile measures how successful people holding a particular position in different organizations have scored on a profile. These are national or multi-organizational norms for managers, supervisors, customer service providers, commissioned sales people, or computer system installers, etc. The information on how successful persons in one of these job roles has scored on a profile then becomes the benchmark for evaluating profile scores of future applicants for those jobs.

Concurrent validity becomes more concrete when it involves developing preferred scoring ranges for a particular position, or a number of positions, within one organization. This can be done in a number of ways, but the most effective way for generating suggested ranges for a job in an organization is conducting a normative study. Normative studies are done by identifying high performers in a position, profiling them, and using those results as benchmarks for profiling scores of future applicants for that position.

There is a large number of personality profiles or tests being used in organizations: the Ghiselli Self Description Inventory, Calipers, and the Minnesota Multiphasic Personality Inventory (MMPI), for example. The MMPI is more of a psychological measure, traits such as paranoia and hypochondria. The distinction between psychological tests and personality profiles is important because in most cases we are looking to measure practical aspects of how identify is related to performance rather than an applicant's psychological abnormalities.

There are a number of factors to consider when deciding on which of the many personality profiles to use. These include cost, whether the profile measures practical characteristics related to job performance, construct and concurrent validity, and efficiency of administration. In addition, it would be best if the profile selected is valid for both outside applicants and persons already working in the organization who are being considered for new job assignments. Of all the instruments available, we have found that The Achiever best meets these requirements.

The Achiever measures six mental aptitudes, ten practical personality characteristics, and has two honesty measures. The mental aptitudes includes mental acuity or quickness to learn and reasoning ability, command of vocabulary, and mechanical interest, among other factors. The personality section of the Achiever has measures of dominance, motivation, extroversion versus introversion, work habits and similar dimensions.

The Achiever also has two honesty or "faking scales" to determine if the person completing it is being honest in answering questions about his or her personality. Specifically, the instrument measures distortion and equivocation. About ten percent of those completing the Achiever, in our experience, distort or equivocate to the extent that the accuracy of the personality results is in question. By inference, a high distortion or equivocation score probably indicates that the applicant has difficulty with feedback. Receiving feedback is part of most jobs.

As mentioned earlier, finding an effective match between the applicant and the position is beneficial to the organization and the applicant. Unfortunately, many organizations conduct hiring in such a way that they lose a large number of their new hires within a short period of time.[8] In many cases, poor hiring results from one or the other of two causes: the hiring process has been poorly defined or is not followed, and decisions on hiring are made without sufficient information. The following case will illustrate more specifics regarding effective hiring and includes one organization's hiring process.

Organization: Training function of a consumer products company

- Markets/Internal Customers
    - the company provides pet food
    - the Training Function develops and provides training in many areas, including management, selling skills, and teamwork
- Structure
    - there were a number of human resources development units, basically organized around the groups of internal clients they serve
- Primary Issues/Problems
    - lacked an effective hiring system: high turnover and not meeting sales goals.
    - needed to hire field sales forces to maximize selling ability and retention of sales persons

The Training Function and their internal clients recognized the need to maximize effectiveness in hiring their field sales force serving the veterinarian and pet specialty stores. Identity factors including personality and attitude directly affect sales performance. In addition, there are specific skills critical to selling success. The manager of training and development for this division of the company developed the selection process illustrated in Figure 6.2.

Selection process

The Training Function realized that the performance of the sales force is essential to success of the unit. The performance of the unit is largely dependent on the field sales force selling in sufficient quantities. An internal pre- and post-study of retention demonstrated that both retention and selling volume was significantly better after the installation of the selection process defined above compared to selling performance before the process improvement. Retention was essentially eliminated and selling levels were increased to reach the goals.

C. Other techniques for effective hiring

Because of the importance and complexity of selection, organizations often make use of additional techniques beyond those identified so far in this chapter.

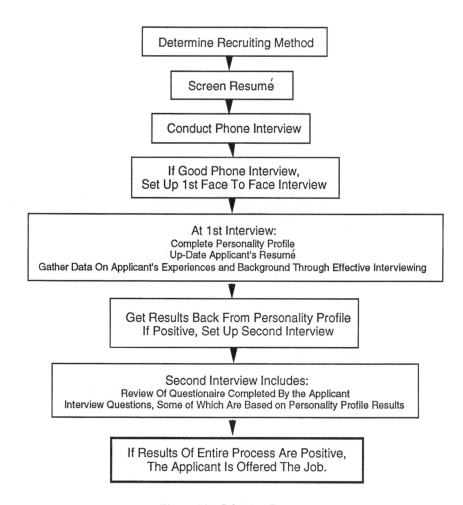

Figure 6.2   Selection Process.

1. Assessment Centers are very common in larger organizations. Depending on the organization, assessment centers perform some combination of the following functions.

- They are involved in the selection and development of employees with high potential, especially with managerial and executive positions.[9] This focus has resulted in some of these units being called "Managerial Assessment Centers."
- Assessment Centers sometimes function as career development centers, although this is probably less common than the selection and development function. The essence of career development is to clarify

the goals and major identity characteristics of the employee and link those aspirations to the objectives of the organization/team in which the employee works.

- The third, probably least common use of Assessment Centers is in choosing and developing people for new work designs, such as teamwork, project matrix structures, and the current emphasis on self management. This often includes career self management.

These three primary functions of Assessment Centers have a common purpose, identifying and improving the fit between candidates and positions. The techniques frequently used by Assessment Centers include:[10]

Objective tests of personality, mental aptitudes, attitudes, and interests
In depth interviews
Individual presentations
Management games, simulations of work experiences, outbound exercises such as wilderness camping, in-basket activities
Feedback on leadership style, teamwork skills and discussion of desirable models for these roles

Assessment Centers are in large part an attempt to systematize the selection process and make sure it is coordinated and thorough. Many organizations have the rudiments of Assessment Centers, even if they do not use the organizational label. Selection in general and approaches like Assessment Centers should be done with an eye to specifically serving the organization's strategy. More about that in the following chapter.

2. Realistic Job Previews is a selection technique whose benefits have been well documented. Not suprisingly, hiring officials having conversations with valued candidates tend to "over-sell" their organizations. They do not describe the negative aspects of the job, the team/department, or the organization. This unrealistic perspective has led to occurrences of "job shock" when a new employee starts a position and discovers the down side.[11] To overcome this problem and the resulting poor morale, organizations sometimes structure an activity where employees are introduced to the negatives of the job, such as boredom, long hours, etc. Among other benefits, this technique appears to reduce turnover.

3. Various other techniques to improve selection exist. These include an applicant experiencing problem solving situations with current managers, or spending time with another employee doing the work they will be doing if selected for the job. These techniques all have the same purpose, providing information which will help all involved in deciding if the fit between candidate and position is mutually beneficial.

D. Three additional selection issues

1. As mentioned earlier in this chapter, selection officials in organizations frequently do a poor job at making transfer or promotion decisions for current employees. The internal politics of the organization, such as whether a person is "in" with managers having influence, or whether the current employee has been "loyal" to the leadership, often has undue influence in making promotion decisions. A candidate from inside the organization should essentially be evaluated the same way as a candidate from the outside. Proper fit of the inside or outside candidate to the position is the most important consideration. While loyalty and good work of current employees should be rewarded and recognized, it is not in the best interest of a good employee to be moved to a position for which he or she is not suited by personality, attitudes, interests or motivation.

2. As a practical matter, the manager or supervisor responsible for directing the candidate should have at least equal influence in the selection decision. Following this principle is hard for many executives or owner/business operators because of the view that leaders should be responsible for getting the right people into the organization/team. The basic point is that the person who will direct the candidate will feel more responsible for helping that person perform well if they have at least equal influence in the selection decision. Managers/supervisors often resent having employees imposed on them by their superiors, and even supervisors who are well intended struggle with this resentment.

3. The selection process is a major opportunity for gathering information that can provide the basis for a number of factors pertaining to the selected candidate: how to manage the person, identification of needed training and development, and what information should be emphasized in the orientation. Specifically, the selection process should provide the following information about the selected candidate:

- Information from profiling, attitude tests and interviewing pertaining to the person's identity; personality strengths and challenges, motivation, values, and basic attitudes.
- Information from interviews, skills tests and reference checks to help the organization/team know what knowledge and skills need to be developed in the new employee. This should lead to plans for technical training as well as knowledge and skills for the person's role as manager, sales associate, customer service provider, or team member.
- Checking references and interviewing, along with feedback from Assessment Centers, provides information on areas of the management style of the new employee which can result in development plans.

A technique beneficial to the organization and the person selected is the development of a hiring agreement as a part of the selection process. This contract should include:

- agreement on basic areas of responsibility and the output require-ments for the newly selected person.
- agreement on training and development programs and activities the person is to experience. This should include listing which specific concepts and skills need to be mastered.
- agreement on the support needed by the selected person in order to maximize performance and what manager(s) will do to provide that support.

E. Conclusion and summary

Selecting the right people for positions in the organization/team is a key to performance excellence. All the performance improvement interventions discussed in this book are less effective if the occupants of positions involved in those efforts are poorly fit for the jobs they hold. On the positive side, the performance interventions described in this book will have major impact when selection has been well done.

F. Suggested action steps for organizational/team leaders

1. Use the information in end note 3 of this chapter to decide on job requirements for areas of hiring/selection needs in your organization.
2. Review the organization's selection process, both for outside appli-cants and internal applicants.
3. Find or develop methods for effective interviewing and make it avail-able to hiring managers.
4. Review available personality profiles, attitude/interest surveys, and skills tests. Install the appropriate instruments at the proper stages in your selection process.
5. Consider conducting a normative study for the most critical positions in your organization/team. Install the "suggested scoring ranges" in your profiling process that result from that study.
6. Decide if any other selection techniques are needed for improving the probability of getting the right people in the right positions.
7. Make sure that your organization/team's promotion and transfer process makes a systematic use of available data sources.

End notes

1. Dessler, Gary, *Personnel/Human Resource Management*, Prentice Hall, Engle-wood Cliffs, 1991, 172.
2. Aldag, Ramon J. and Stearns, Timothy M., *Management*, College Division, South-Western Publishing, Cincinnati, 1991, 392.
3. We discuss individual job competencies in Chapter 10. For a thorough dis-cussion of the subject, see Dubois, David D., *Competency-Based Performance Improvement*, HRD Press, Amherst, 1993.
4. These guidelines are a summary from an A.A.I. training program on hiring.
5. Dessler, Gary, *Personnel/Human Resource Management*, Prentice Hall, Engle-wood Cliffs, 1991, 173.
6. *SRA Human Resources Assessment Catalogue*, Pg. 4, and the *SRA Test of Mechanical Concepts Examiners' Manual*, London House Products Group, Rose-mont.
7. Dessler, Gary, *Personnel/Human Resource Management*, Prentice Hall, Engle-wood Cliffs, 1991, 180.
8. Stoner, James A.F., Freeman, R. Edward, *Management*, Prentice Hall, Engle-wood Cliffs, 1992, 562.
9. Cummings, Thomas G. and Worley, Christopher G., *Organizational Develop-ment & Change*, South-Western Publishing, Cincinnati, OH, 1997, 411.
10. Dessler, Gary, *Personnel/Human Resource Management*, Prentice Hall, Engle-wood Cliffs, 1991, 189.
11. Aldag, Ramon J. and Stearns, Timothy M., *Management*, College Division, South-Western Publishing, Cincinnati, 1991.

chapter seven

Strategic plan for the organization: Where it all starts

A. Connecting strategy to prior discussions

Previous chapters have discussed various "strategic" concepts, elements and benefits of thorough and well communicated strategy. A brief summary of previous points will help readers make the transition to this chapter.

1. In Chapter 1 the point was made that many organizational/team leaders feel a powerful need to engage in performance improvement, but often the motivation to improve performance and the selection of programs used is triggered by mundane considerations such as the "latest and greatest" performance intervention. Chapter 1 ended with an introductory discussion of the role of the strategic vision of leadership in deciding the direction and nature of the organization/team and therefore what type of performance is needed.

2. Chapter 2 took the concept of strategic vision and direction further, indicating that the ideal situation is that the leadership develop a thorough and specific strategic concept, major objectives to accomplish that strategic concept and communicates it to the rest of the organization. What occurs then is that levels of the organization/team below the leadership begin to understand overall direction of the organization and can set goals, standards and models of effective performance for their own areas. This permits the measurement of current performance and the ability to identify performance gaps between desired and actual performance.

   Unfortunately, in the real world, getting leadership to clarify the strategic direction is often the most urgent need because it has not occurred in the past.

3. Chapter 3 discussed methods for deciding on performance improvement, whether it is needed, and if so, what interventions are best. One major type of performance intervention, as demonstrated in the Organizational Success Model (Chapter 3), is strategic planning. Other performance improvement interventions are most effective when they are seen in the light of the overall strategic direction.
4. Chapter 4 discussed goal setting and Key Performance Indicators, these goals and measures having "cascaded" from the strategic vision of the leadership.
5. Chapters 5 and 6 discussed two fundamental features and processes of successful organizations: creating a learning organization, and selection processes for filling positions with outside or inside applicants.

Learning organizations use their strategic vision as a driving force for their learning curriculum, and part of their learning curriculum is the strategic vision of the organization.[1] Hiring affects strategy because the direction the organization/team takes is influenced by the knowledge and skills, the experience and motivation, and even the personalities of the people in the group. What is less clearly understood is that strategic direction should also help determine what knowledge and skills, experience, behavioral styles, and basic values should be emphasized in selection of new employees.

B. Process for creating organizational/team strategy

The strategy of an organization, or an independent team, is a thorough and detailed statement of what the leadership desires and intends the organization to be. It is the organization at its best, as defined by the leadership. The following is a description of what the strategic plan includes. The next section of this chapter discusses a process for developing the strategic plan.[2]

I. The plan begins with two essential decisions, the strategic time frame and identification of the organization's driving force.

The strategic concept statement, discussed below, is a written statement of the desired and intended future of the organization, a statement about a future set of conditions for the organization. At least to some degree, the organization/team is not currently everything the leadership desires or intends. Therefore, the leadership group creating the strategic plan need to pick a point in the future, usually three to six years ahead, to build its future strategic vision. As organizational leaders often say, "it takes time to get there." That time is the strategic time frame.

There is no objectively right or wrong date for the strategic time frame. Usually organizations pick the date that marks the beginning of their fiscal year. Other than that, there are only general guidelines. A strategic time frame of less than three years is too short because it becomes more like long range planning or budgeting, two years as usual put back to back. More than six years is usually a problem for United States business leaders because of their tendency to focus on short term operational problems. Decision makers choosing a strategic time frame often use considerations of general length of life for product lines, churn or turnover in major markets, and how long the CEO has before retirement. Other considerations that occur are rate of technological change and availability of capital.

The driving force is a much more complicated decision and one where the leadership team needs to make the right decision. Some basic points need to be made about driving force.

- In large part, the essence of any work organization is the products and services it provides and the markets it serves. The rest of the organization (technology, marketing approach and techniques, financial and human resources, structure and culture) support the creation, marketing, and delivery of the products to customers. Thus, the most fundamental strategic decision is: What should the scope of our products/services and markets be?
- Determining the scope of the organizations products/services and markets affects the entire nature of the organization. If we make a strategic decision to offer a new product line, or drop a current one, we will be making decisions affecting human resources, marketing strategy, and production. If we decide to begin marketing our products in a different country, or even a different state, at minimum we have made decisions impacting the distribution systems we use and the marketing reach we seek to accomplish.
- The driving force is the single strategic element which determines and drives the organization's total strategic concept. The driving force dictates future products and services; it is a major determinant of strategic decisions and future resource allocations; it is the central theme and focus of the organization/team. The choice of driving force is a conscious decision made during strategic planning.
- While there are a number of driving forces possible, products/services offered and markets served are by far the most common.

1. If an organization uses products/services offered as its driving force, it is saying that a well defined array of product lines is the central focus of its organization, and it will spend a good deal of its energy and resources marketing those products and services everywhere it deems appropriate. New product development outside of the basic product lines is de-emphasized, and marketing is emphasized.

In the 1980s, Anheuser-Busch had a wide variety of products and services: adventure parks, wine coolers, snack foods, luxury liners, a major league baseball team, and, of course, beer. At some point the leadership made the strategic decision that the beer products were to become the focus. In our terms used here, they decided their driving force was products, the many brands they were manufacturing. That led Anheuser-Busch to sell off many existing business outside the primary product line.

A few additional clarifying points are important here. People drink beer at baseball games, adventure parks, and luxury liners. But those are "secondary products" in those events, not primary products. As Anheuser-Busch began eliminating peripheral businesses, it also expanded its marketing of the primary product line. The company now sells in Mexico, Asia, Latin America, and much of Europe. A-B also added to its primary product lines. But, except for adventure parks packaging, it is all beer.

2. If an organization uses markets served as its driving force, it is committing to understanding its defined market as well as possible, and then doing what it can within its capability and capacity to provide products and services needed by that market. In this case, the organization spends an unusual amount of time conducting market research and developing products, sometimes even outside of its primary product lines.

Examples of markets served organizations include Gillette Company and Merrill Lynch & Co. Inc. In both cases, their approach is to define and understand their market's demographics and provide a multitude of services to meet that market's needs. An important point of clarification is that Gillette, for example, offers a wide variety of different products: razor blades, deodorant, shaving cream, after shave, etc. The technology for making these products is very different; some are liquid and some metal. What makes them similar is their end users, "the market served."

3. Other options for driving force exist, though they are less common. Technology is the driving force for Texas Instruments. A highly structured distribution system is the driving force for McDonald's Corporation. And a distinctive selling method is the driving force for Avon, Amway, and Spiegel. Capacity is one of the most unusual driving forces, with Moore Business Forms being an example in that they have the capacity to be a major supplier to the Internal Revenue Service.

- Deciding about driving force is probably the most complicated and most important strategic decision for an organization. Many organizations/teams have never thought about making a driving force decision, and so they float between multiple driving forces, "trying to be everything to everyone." The problem with that approach is lack of focus on core business, dilution of resources, and confusion about what opportunities to pursue and which ones not to pursue. An organization, no matter how large or extensive its resources, cannot afford multiple driving forces.
- Changing driving forces can be done consciously, but requires great consideration because of the costs. If, for example, Anheuser-Busch was to decide to become markets-served driven instead of products/

services driven, they would need to define their primary market clearly. If they defined that market in a specific way, for example "beer drinkers in the Americas and other developed nations," the driving force would suggest that they then begin conducting or confirming research on non-beer products and services wanted by that market. The brewery is famous for having excellent marketing management and research capabilities, so they could undoubtedly accomplish the research task. The problem then would be to develop the capacity and capability to produce, market and distribute new products services for their market. One can envision a major shift toward a number of products such as beer tappers and portable pre-fab bars for home and business use, information on food and beer combinations, T shirts and beer mugs. The point is that these are very different products from the production of beer.

A brief case will help illustrate the importance of the driving force decision. This case was used to make a different point in Chapter 4.

Company: Tone's Brothers

- Products and Services
    processed spices
- Markets
    retail chains
    food manufacturers
    cafeteria management companies
- Structure
    plant of approximately 1000 people
    about 200 managers and sales and customer service personnel
    unionized factory workers
- Primary issues
    organization had hired a new human resources director and
        significantly added to the H.R. staff.

The new human resources director, well educated and experienced in organizational performance, realized that he and his team members needed a planned direction. They understood the process for strategic planning described in this chapter. In the beginning of the process, they discussed driving force. After substantial consideration, their choices came down to two alternatives. First, they could choose to be products/services driven. If they did that, they would identify the products/services they thought they could and should provide and undertake developing the capacity and capability to deliver those products and services.

OR

Secondly, they could decide they were driven by markets served, identify those markets, and strive to understand what was needed for each market group.

The team chose the second approach. Their reasoning was as follows. "We know who needs human resources services better than we know what they need, beyond standard categories of services of benefits and some type of training. We have the capacity to develop whatever is truly needed, and we can develop the capability to provide what is needed. We do not have a substantial list of existing services, and have the opportunity to build products/services in response to our internal customers (markets)." The team then identified their internal markets, primarily managerial groups and front line workers. Following that they began to create their Strategic Concept Statement, the second component of the strategic plan following driving force decisions.

Before we discuss the Strategic Concept Statement, a word of caution about the driving force of markets served is useful. Markets, clients, citizens, or customers served (the labels depends on the organization doing the planning) is not the same as customer service. Good customer service is about reasonable credit and product return policy, on time delivery, repair services and friendly and responsive interactions with sales and service people. Markets served as a driving force is using the needs of the market to determine what products and services your company will create and provide.

II. The second component of the strategic plan is the strategic concept statement.

This is a detailed word picture, describing the desired and intended future of the organization/team. It is the word picture, or vision of the unit as leadership would like it to be. The leadership team creating the strategic plan makes the following decisions:

"By this date: _____ (3 to 6 years into the future) what do we desire and intend to have as our primary products and services and our secondary products and services? By this date: _____ what do we desired and intend to have as our primary markets served and our secondary markets served?" The discussion around this question produces a picture or vision of what the leadership wants to provide and to what markets. The same question and resulting decisions are also applied to five other strategic elements: sales method, (or marketing in some organizations, communication in government agencies, etc.); distribution system; technology; major sources; capacity and capability.

What results is a detailed description of the desired/intended organization. It is much more than two paragraphs of "who we are," sometimes called a mission statement or values statement. It is a plan for the future, a model of "effective performance" for the organization. In addition to narrative descriptions of desired and intended products/services, technology and the like, it contains Key Performance indicators for the whole organization/

team which measure the progress towards the strategic vision. Numerical values are established for each year during the strategic time frame and are changed as experience dictates.

Earlier chapters have discussed organizations and independent teams as if they were much alike strategically and in terms of setting direction and goals for team members. If a team has the independence to make choices about its own products and services, markets it serves, and its technology and major sources, then it can develop its own strategic plan. Human Resources at Tone's Brothers discussed above and in Chapter 4, and Ralston Purina's Pro*Visions, also discussed in Chapter 4 and below, are examples of independent teams who benefited from developing their own strategy. The main issue for the strategy of a team within a larger organization is that it must also account for the strategy, desires and wishes of the organization of which it is a part. That is, however, not a great deal different from a total organization being concerned about the strategy, desires and wishes of their major customers.

All organizations have a strategic concept or vision, meaning some more or less clear definition of products and services, markets or clients being served, and the like. But as we have said before, that strategic vision is often vague, contained only in the mind of the top leader, different for each of the top leaders, and communicated poorly. The process of strategic planning is aimed at creating a shared picture of the desired and intended future organization/team.

III. The third component of the strategic plan is the evaluation of the current situation, both internally and externally.

This is where internal Strengths and Weaknesses, and external Opportunities and Threats (SWOT) are listed and discussed. The best approach is to evaluate the current situation items listed in terms of their importance to achieving the strategic concept statement. That is why it is necessary to create the vision of the future before the evaluation of the current situation. A simple three part rating system (1-2-3), with 3 indicating major significance of the situational item, and 1 indicating minor significance, is helpful.

Major elements of the internal situation which are reviewed in light of the strategic concept statement include the following: facilities and equipment; systems, methods and procedures; leadership, management, communication and control; human resources; current products and services, sales and marketing, and financial systems and status. Each area is taken in turn, and the data for identifying items within, for example, financial systems, comes from the leadership developing the strategy. An opportunity for broadening involvement of persons not in the strategic leadership team exists, however. That opportunity is to generate input from other members of the organization on operational items they see as strengths or weaknesses in the SWOT areas.

In most cases employees other than leadership will be particularly insightful about internal strengths and weaknesses. People with customer contact may also see external opportunities and threats in products/services, sales method or technology. Gathering information from others in the organization broadens interest in the strategic plan, usually setting up a good basis for communicating the completed strategy to the entire employee group. The evaluation of the strategic importance of the SWOT items suggested by other employees is done by the strategic leadership team in light of the strategic vision. One of the interesting side benefits of this process is that it drives identification of those SWOT elements all organizations live with. Some things are irritations, others are strategically important, and the process clarifies which is which.

IV. The fourth component of the strategic plan is strategic objectives (goals).

The process for defining these goals is as follows. The *Strategic Concept Statement (Vision)* is where the leadership group desires and intends the organization to be at some point in the future. The *Current Situation* is where the organization/team is currently in a number of areas, and an evaluation of the strategic importance of elements within those areas as described above.

> The strategic objectives are the immediate goals and actions to make progress from where the organization is currently, to where leadership desires and intends it to be. This is where the action is; this is the center piece of organizational performance.

Strategic Objectives are set for immediate action, with people responsible for making them happen somehow identified. Preliminary action steps should be created by the strategic planning leadership team because they have created a shared vision of the future, and action steps to make progress in that direction are usually quick to flow from the leadership team. The key is that each Strategic Objective have a champion, someone accountable for making the objective happen. Of course that champion may need a number of people to help in achievement of the objective.

Figure 7.1 is a visual description of the Strategic Planning Process discussed above.

C. Cascading goals revisited and extended

At numerous places earlier in this book I discussed the importance of using the strategic vision of leadership to cascade goals throughout all levels of the organization team. After the strategy is developed, this should start with a presentation to all of the employees of the basics of the newly developed strategic plan. In general, communicating the primary elements of the Strategic Concept Statement (vision) and the Strategic Objectives, is the most important. The details of the SWOT evaluation are more than is needed,

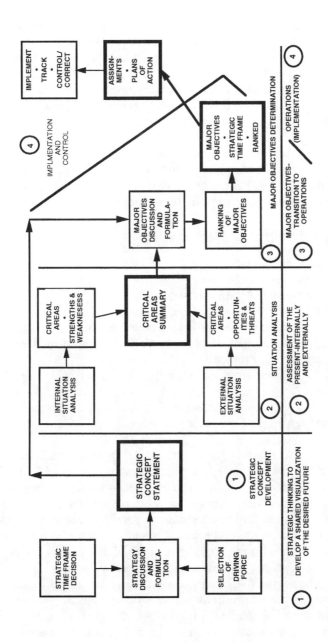

Figure 7.1   Strategic Planning Process. (From Weaver, W.C., Achievement Associates, Inc., St. Louis, © 2000. With permission.)

although strategic items within each strategic element which have been rated as highly significant to strategic achievement should also be discussed with the leaders of the unit directly affected. For example, if the SWOT analysis determined that current communications equipment in the organization was deficient, budget and purchasing personnel at least, should be informed.

Experience shows that the single most important reason why organization/teams fail to make a strategy happen does not come from weaknesses in the plan itself. Weaknesses can be overcome with strategic leadership team meetings on a regular basis where review of progress on Strategic Objectives is conducted and plans adjusted. Rather, the single most important cause of strategic failure is due to lack of persistent efforts at achieving the Strategic Objectives.

There are a number of ways to manage actions at achieving Strategic Objectives. The most common is for the CEO of the organization to manage the actions through his/her top team. Two brief case studies will clarify this. The first is a case discussed originally in Chapter 4.

Organization: Scottsdale Securities, Inc.

- Products and Services
    on-line discount broker
    broker-assisted trades as needed
- Structure
    approximately 100 central corporate personnel
    approximately 106 branches spread throughout the United States
- Primary Issues/Problems
    lacked shared strategic vision
- Actions
    conducted a thorough strategic planning process with twenty or
        so managers at the corporate level

The CEO had a large number of direct reports. He and his corporate department leaders, engaged in strategic planning which resulted in 19 Strategic Objectives. These objectives covered a number of strategic areas, creation of a few major new products/services, developing new technology which is a foundation for on-line brokers, and human resources issues from creating a training function to customer service improvement in the branches. The CEO managed the actions on these objectives by having persons responsible for each one develop and communicate action plans and steps. He then conducted meetings with the group as a whole to discuss progress on each objective and to give the group time to know how objectives other than their own were progressing. While the meetings were not as frequent as the CEO and his top advisors wanted, the process of managing the achievement of these objectives from the top down worked. Within approximately one year of the creation of the strategic plan, all but about 5 of the 19 major Strategic Objectives had been completed or there are been significant progress on them.

In today's fast-paced world it is almost always true that pressures of daily operations interfere with the long term performance improvements that come from a strategic plan. This is even more true when the leadership of the organization is more complicated than at Scottsdale Securities, Inc., with a single CEO who is also the major owner. The following case illustrates challenges from multiple ownership, but also shows how even separate leadership issues can be partially overcome with a strategic plan.

Organization: Roberts, Perryman, Bomkamp & Meives, P.C. a mid-sized law firm

- Products and Services
     litigation: personal injury for defense
     business legal services
- Markets
     casualty and property insurance carriers
     small- to medium-sized businesses
     any self-insured organization within their geographic market
- Structure
     headed by six partners, with a managing partner who was also a
       major owner
     approximately 10 additional attorneys
     approximately 20 other staff from paralegals to clerical support
- Major Issues and Problems
     the managing partner believed the group lacked commonly agreed
       on direction and had major disagreements over marketing their
       services, financial direction, profit sharing and hiring decisions.
- Actions
     conducted strategic planning with the partners

Enough other partners agreed with the managing partner's perspective on the organizational issues that they decided to engage in strategic planning, involving the six partners. After a lengthy process and many discussions, the attorney partners agreed to five major objectives including a marketing/sales plan, major technology improvements, financial/procedures and systems development, and a legal skills development program. Two of the five Strategic Objectives were accomplished quite quickly; three were slowed by disagreements between the partners. But persistence on the part of three of the partners helped with significant progress on the final three objectives within 18 months of the creation of the strategic plan. As of the writing of this chapter, the law firm has updated the strategic plan and added new objectives.

Some organizations manage actions for achieving the Strategic Objectives through cascading those objectives to teams and making the team leaders responsible for their accomplishment. Here the role of the top leader is somewhat less critical. This approach works best when the organization/team is made up of distinct groups having their clearly defined functions. Again, a brief case will clarify this.

Organization: Webster University, Development Department

- Products and Services
    graduate education
    undergraduate education
- Structure
    organized into three teams by function and client group
- Primary Issues and Problems
    needed a strategic plan with objectives that could help insure that
        each of the teams was working together and in the same strategic
        direction
- Actions
    developed a strategic plan with 14 Strategic Objectives

The strategic leadership group developed a strategic plan and then made each of the three top team leaders individually responsible for moving the objectives ahead. By the end of the first year after development of the strategy, better than half of the Strategic Objectives had been completed. While the top manager was involved in providing advice and assistance to the team leaders, they were each principally responsible for completing those objectives themselves with support from people within their own team.

A continuation of the cascading of Strategic Objectives from top manager to team leader to individual action plans within each team is the final step in using strategy to define performance. This also increases the probability that the objectives will be accomplished. The ideal is for the team/department leader to take responsibility for making sure the objective is accomplished but doing that by having the appropriate role for each person within their group included in the action plans for the strategic objectives. The action plans for each individual should also include appropriate goals for achieving strategic objectives. In short, all the individuals' strategic action plans and goals taken together comprise the strategic goals and action plans of their team and all of the various department/team strategic action plans comprise those of the total organization.

D. Major process issues

Any major effort at performance improvement, of which strategic planning is an example, requires dedication and persistence. Two process issues often occur jeopardizing the effectiveness of strategic planning.

1. Hard work and persistence is needed. The effort at *developing* the Strategic Concept Statement requires vision and creativity, as well as open communication. Also the effort at *evaluating* the current situation requires persistence, determination, and continued openness in communication. When current situation elements are evaluated and many are identified as major weaknesses, participants in the process can be sensitive and defensive, diluting the current situation analysis.

2. Deciding who should be on the strategic leadership team is often a complex issue. Two principles are important here. First, anyone should be included who is important in deciding the basic strategy of the organization/team or critical in making it happen. Secondly, any exclusions should be done on a rational basis, not a "political" one. Those expected to be included but were not should have an understanding of why they were not selected to be in the strategic leadership group.

E. Conclusion

Developing and communicating the strategic direction of the organization/team is the first obligation of leadership. The second is making sure it happens. Communicating the strategy to the entire group is a start on making performance happen. Having an organization which is committed to learning and having a process for hiring good people are both foundations for achieving strategy. The rest of Part II of this book discusses other performance improvement interventions to help the organization/team perform successfully.

F. Suggested action steps for organizational/team leaders

1. Make sure you understand what a strategy is and what it is not. Distinguish strategy from yearly operational plans, value and mission statements, and short term goals.
2. Determine if the strategy is commonly understood by your people in detail and accepted by the team of persons critical to making it happen.
3. If the strategic direction is commonly understood and accepted, make sure you have a process for regular meetings with those responsible for achieving the strategic objectives. Recognize that there are also daily operational issues required of all leaders, but the strategic objectives should have deadlines nonetheless.
4. If there is insufficient agreement on strategy (e.g., products/services/markets served/marketing and selling system, etc.), then get your leadership group together to go through development of a strategic plan.
5. Make it happen.

End notes

1. Meister, Jeanne C., Corporate Universities: Revised and Updated Edition, ASTD, McGraw-Hill, New York, 1998, 93.
2. Many of the concepts for the strategic planning methodology discussed in this chapter were influenced by a class text in the field. Our thanks to Tregoe, Benjamin B. and Zimmerman, John W., *Top Management Strategy*, A Touchstone Book, Simon & Schuster, New York, 1980.

chapter eight

Teams and performance

A. Linkage and overview

For many years the academic field of organizational behavior has studied teamwork with particular interest in the performance of teams versus individuals.[1] In general, it is well accepted that teamwork, when done honestly with real issues, increases acceptance of team actions with the team members, improves speed and accuracy in decision-making, and probably improves productivity in some situations.[2] Self-managed teams, those with control over areas of scheduling their work, hiring into their team, and recognition and rewards, also tend to increase productivity and motivation.

Teams, like any other area of human performance, are not an automatic quick fix. When used the right way and at the right time, they have fairly predictable impact of synergy in decision making and improved productivity, often because of anxiety about letting the team down. As mentioned above, there is also greater acceptance of team decisions if authority to make the decisions is real.[3] Based on the recognition that these benefits of teamwork are valuable, we have recommended the use of teams at various points of performance improvement.

B. Some areas for teamwork

1. The development of the Strategic Plan for the organization/independent team should be done by a leadership group, not by a single individual. The benefits are obvious and are summed up in the statement that no single individual can make an organization perform successfully, except in a single person proprietorship. In fact, it takes a collection of inter-dependent teams and/or individuals to make a strategy happen. Synergy in planning and execution, acceptance of the plan and motivation to accomplish the strategic objectives is usually greater for those involved in the development of the plan, compared to those handed the plan by their leadership.

2. Management of the actions toward achieving the strategic objectives will be improved through a team approach, whether this is the top team under the direction of the top manager, or teams throughout the organization charged with achieving "their part" of the objectives. Motivation to keep going on strategic objectives, and creativity in deciding on actions, are all heightened by a good team.

3. Significant decisions about hiring should be done in a team setting. Remember that the more information about a candidate you have, the better the decision is apt to be. Input about the decisions from all persons significantly affected by the selection is valuable. Also, interviewing is affected by the attitudes and background of the interviewer, and balancing and contrasting one interviewer's perspectives with others can make the interview data more objective.

4. At the operational level, getting teams involved in setting their shared goals, whether strategic or operational, usually adds to a commitment to performance in achieving those goals compared to people doing individual action plans in isolation.

As mentioned earlier in this book, strategic objectives are focused on the longer term development of the organization/team with the purpose of continuously moving toward a detailed vision of excellence in performance. But daily operations require goals that are shorter term, today, this week or this month. Examples include making an important sale, solving the immediate technology problem, fixing the machine on production line four, and getting the new bank loan. While these immediate goals should serve (or at least be consistent with) longer term strategic goals, the pressure of today's activity is real and important in keeping the organization/team functioning. Teamwork on a daily basis, working to activities to achieve immediate operational goals is also important to successful performance. This chapter will discuss the importance and limitations of daily team work in supervision and production, marketing and selling, and customer service. Chapter 9 will discuss the leadership team.

C. When to use teams and when not to use them

In Chapter 1 we discussed the power of fads, what is popular in performance improvement becoming "the thing to do" for organizational leaders. A specific example of the power of fad is teamwork. Many people automatically jump to the conclusion that teamwork is what is needed in many situations. While the benefits of teamwork are real, as we have indicated above, there are times to use teams and times when teams are not needed. The model in Figure 8.1 helps clarify guidelines for team use.

The two dimensions of concern in this model are the extent to which a given decision or project requires decision quality and when the issue is decision acceptance. As indicated above, depending on the project or topic

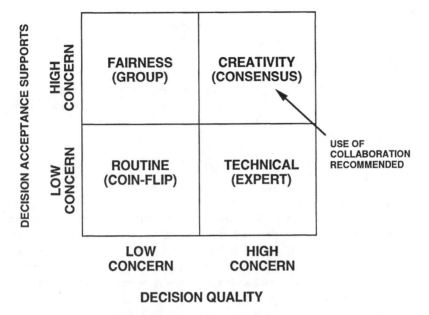

Figure 8.1 Decision Process Tradeoffs.

involved, teams can improve quality in decisions, as well as acceptance. But teams are not always needed. Here is how it works.

Some decisions fit into the upper left hand quadrant; there is high concern for acceptance, but low concern for the decision quality in the sense of the decision's importance to the organization/teams performance. One example which occurs frequently is decisions about health care policies provided to the employees. Many organizations will find a small number of health coverage programs they believe are roughly the same in terms of costs to the organization, administrative responsibility, etc. The decisions about which of the alternatives to make available to the employees in these cases has more to do with the personal lifestyle and attitudes of those employees. Acceptance and fairness is the primary consideration. In this case, a survey of employee desires may be the best approach to making the decision among the options the organization has developed. A team decision is not the best approach.

The quadrant in the lower left hand corner involves the daily routine decisions that involve little concern about the quality of the decision and little concern about its acceptance. The decision has to be made, but no one really cares except the person stuck with the responsibility for deciding. Decisions about routine office supplies or what color to paint the restroom are examples. Somebody has to decide, but assuming some good judgment on the part of the decision maker, it's their choice. Again, no team is needed.

The lower right hand quadrant is more complex than the first two. Here the concern is with the quality of the decision, but its acceptance is much less of an issue. The decision is technical, requiring an expert. An example here is the asset investment program of the organization, usually a decision made by the CFO with some direction from the CEO.

Finally, we get to situations where teamwork is needed, the upper right hand quadrant. Here your concern is both with the quality and the acceptance of the decision. Examples are developing the strategic plan, setting departmental/team goals, decisions in major projects and significant hiring decisions. This is where effective teamwork is essential.

D. A model of effective teamwork

Virtually everybody who understands work teams agrees on a few basic concepts. A team is a group of people who are working together to achieve a goal or number of goals. There are different types of teams: permanent teams such as marketing, production, or sales teams; project teams with a specific assignment and usually a limited life; and top leadership teams charged with providing leadership. These are the most common examples. A less common team is the "self-managed" or autonomous team, used where responsibility for the entire product or service can be vested in a single group.

Experience shows that there is a model of effective teamwork, regardless of what type of team it is or what it does (Figure 8.2).

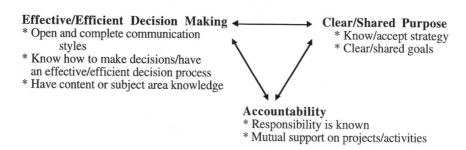

Effective/Efficient Decision Making ⟷ **Clear/Shared Purpose**
* Open and complete communication styles
* Know how to make decisions/have an effective/efficient decision process
* Have content or subject area knowledge

* Know/accept strategy
* Clear/shared goals

Accountability
* Responsibility is known
* Mutual support on projects/activities

Figure 8.2   Model of effective teamwork.[4]

The model in Figure 8.2 is logically obvious, but a brief discussion will help clarify it. To be effective a team has to know what it is seeking to accomplish. Remember that clear goals, strategic objectives, standards, and Key Performance Indicators, set the basis for performance. Without this clear/shared purpose, no one knows specifically what performance is or how the team is doing currently, and therefore there is no rational basis for performance improvement. Again, strategy is in large part the products/services the team provides, and who they provide them to (internal or external customers).

While Clear/Shared Purpose, the beginning point for team building, is important, the team also has to be effective and efficient at decision making. That means the team has to make good decisions about its purpose and what actions to take to perform well in accomplishing its goals. This requires three things: open communication and trust, having a process for team decision making that works well for the team, and knowing the content or subject area in which decisions are made.

The third component of effective team work, accountability, is an element frequently missing in teams. Absence of accountability is caused by the same factors as absence of goal setting. People often fear that if they accept accountability for doing something, accomplishing a goal or helping the team perform better, they will be punished if they fail. So they choose to avoid accountability where possible. The cause for this sad condition is not primarily weaknesses on the part of employees. It is ineffective leadership style. More about that in the next chapter.

The three components of leadership reinforce each other. If the team has clear purpose, discussing what to do to achieve those goals or strategy has focus. But also, defining purpose is part of what teams need to decide. If a team has been assigned responsibility, what is needed is clarifying responsibility. So, clear purpose helps with team decision making, and team decision making has to start with clarification of purpose. It is troubling to see how many teams spend a good deal of time in discussions without asking the following question; "What are we trying to do here?"

Honesty in team discussions about purpose or the actions needed to achieve the goals, is frequently so strained that the benefits of synergy do not occur, and people on the team do not accept or "buy into" the decisions that do occur. The most common example of this is a team dominated by a strong personality, or a person in the position of greatest power, like an owner or CEO. Leadership has a lot to do with how well the team functions. If leadership sincerely encourages openness in communication, it is apt to happen. If leaders are threatening when communication does happen, it will stop.[5]

Clear purpose and goals also impact accountability. No one can be held accountable if the goals or purpose are not clearly defined and accepted. But even if team goals are clear, identifying who is accountable for taking action to make each goal happen is essential. How many times do team members go away from meetings asking, "Who did we say is responsible for that?"

The relationship between accountability and team decision making in the model of effective teamwork is as follows: the team decides who is responsible to do what in terms of specific goals. In addition, every team member is responsible for making sure that decision making in the group is effective. Team members must be willing to ask questions about honesty in the group and effectiveness of team decision making. In short, they are accountable not only for their part of the team goals, they are also accountable for helping the team, including themselves, function effectively as a group.

Since a team is a group of people functioning together to accomplish a common goal or goals, cooperation between team members is needed in many situations. In fact, it is needed any time two or more people are sharing work or have different parts of the work process. Also, inter-team cooperation and coordination is frequently needed and yet difficult to accomplish at times. If team A performs steps 1 and 4, and team B does steps 2 and 3 in the work process, then inter-team cooperation is needed. Despite the obvious need for inter-team coordination in these situations, it is often deficient, and the deficiency can be caused by factors such as physical separation or personality conflicts.[6]

Inter-team coordination and cooperation can be accomplished essentially the same way as teamwork within one group. The key is to develop or clarify common goals between the two groups. Following that, the groups need to learn to communicate with one another to work at actions for accomplishing their common goals. Ironically, sometimes the stronger the teamwork within an individual team, the less willing they are to cooperate with other teams. Still, whatever the reasons for lack of cooperation or outright hostility between two or more teams, actions can be taken to improve their working together. The ACF case below will help clarify this point.

It is probably obvious that team cooperation is essential in a leadership team developing strategy, or reviewing how strategy has been going. It is also needed in project teams and other temporary groups. It is perhaps most critical in more permanent functional groups like those engaged in marketing, supervision and production, or selling. Because of the on-going nature of the permanent team members working together, effective teamwork is highly beneficial to performance, and poor teamwork is disastrous. What causes so many teams to be so dysfunctional? Since human beings do appear to have a natural tendency to form groups, why are so many groups ineffective at achieving goals?[6]

There are multiple reasons for ineffective teamwork. Sometimes the team does not know what they are doing wrong together or how to fix it. The Model of Effective Team work can help there. Sometimes they do not have the subject matter expertise to make good decisions, so they need either to be educated or bring in subject matter experts. Sometimes they honestly disagree on basic goals or purpose, then they need to develop a strategy. As often as not, there is a personality conflict going on between team members that make openness in communication or working together on goal achievement, more difficult.

Highly assertive personalities often dominate or suppress communication in team meetings, particularly if the power of personality is reinforced by positional power. But two highly assertive personalities can be even more of a team problem, as the two dominant forces struggle over control and power within the group. On the reverse side, if all the members of the team lack assertiveness, the group can be so reserved as to be lethargic. The right balance of personalities in a working group is possible to attain, but not easy. The following cases will help clarify and extend some of these points regarding teamwork.

Organization: ACF (American Car Foundry)

- Products and Services
    manufactures and leases or sells railroad cars
- Markets
    companies who depend on rail to transport their products
- Structure
    a single owner who sets major strategy but does not directly
      manage the company
    corporate office separate from manufacturing plant
    plant in West Virginia is unionized
- Primary Issue/Problem
    corporate staff had evidence that communication between plant
      management and union leadership was especially poor, and
      hostility between the two groups was impacting performance in
      negative ways
- Actions
    find a way to build at least temporary cooperation between union
      and plant leadership on some critical issues

This is a classic example of inter-team conflict. It is also structurally caused because the two groups have very different goals and separate legal existence. Hostility was at an all time high. The following practical issues were occurring: there was a major safety issue that the groups needed to solve but were having trouble doing so. The backlog in grievances was unusually high and growing. There was major disagreement between plant management and union leadership over "contracting out," the process of hiring external laborers to perform work in the plan deemed to be outside the capability or capacity of full time employees. Finally, there were hints of concern over quality of the product because both groups realized that the quality of their products was related to the ability to compete, keep customers happy, and stay operational.

The intervention involved conducting an efficient survey on how each group saw communication, decisions, and other aspects of the organizational culture. The data was presented on how each group saw the problems and opportunities for improvement and how the two groups combined saw these issues. To the surprise of labor and management, the two groups saw the issues and opportunities in virtually the same way. That set the basis for the possibility of inter-team cooperation.

The second step in the intervention was to have each group separately define the three major issues clarified by the survey (safety, "contracting out," and grievances) and identify a fourth issue as well. Both groups identified quality of their product as a major issue of concern. Each team, labor and management, also separately made decisions about actions to take in solving or making progress on these problems.

The final step was to get both labor and management leadership together and discuss the goals, actions, and problem definitions that had been developed separately. The commonality of how the two groups saw the problems and desirable actions to solve those problems was striking. Combined teams of labor and management then were assigned the goals of taking the input from both labor and management and creating a common action plan. One team worked on safety, one team on "contracting out," and one team each on the other two issues. Each of the three "merged" teams developed action steps despite their suspicion regarding each other. Within one year of this intervention, the safety problem had been solved, major progress had been made in reducing the occurrence and backlog of grievances, and some progress had been made on development of a mutually acceptable process for contracting with external labor. Only a small amount of progress was made on the quality of progress problem, partly because it was the most complicated program these merged teams developed.

The development of common goals between the two antagonistic groups helped them make progress on immediate issues. Personality conflicts did play a role in the inter-team problem solving, as they usually do. These issues are always present and can only be softened by keeping those in conflict focused on the common goals and purpose. The phrase which works best is this: "I don't have to like you, but we are in this together, and our mutual benefits dictates that we need to work together."

Sometimes personality clashes, or other conflicts of basic disagreement on purpose are so strong that efforts to build teamwork do not succeed, even to the extent that occurred at ACF. This is when reconstituting the team may be required. However, as often as not, attempts at building an effective team suffer from lack of persistence on the part of the leadership of the team. Team leaders have "accountability" for development of the team's commitment to their common goals and being effective at decision-making. Because of their position, team leaders have the potential for greater impact on the team, for good or bad. The single most important factor in how well the group functions is the attitude of the leader toward the team.

E. Marketing, sales, and customer service teams

Marketing, sales and customer service teams are constituted in many different ways and have different responsibility in different organizations. But a few basics are very common. First, sometimes selling is primarily order taking, as is usually the case in e-commerce, but even then, effective marketing, selling and customer service are involved to some degree. Second, when marketing, selling, and service are in the same unit, as was true in the first case presented below, a clear definition of accountability is sometimes a casualty. Third, when they are separated organizationally, inter-group teamwork often becomes the problem. Both organizational arrangements have challenges to effective performance that must be confronted.

For clarity in a frequently confusing situation, let's say that basically marketing involves product and market planning and research, packaging, promotions, price considerations, and usually decisions about which market "segments" to approach. One corporate entity listed the following areas as central to the performance of marketing associates.[7]

- Business acumen
- Customer focus
- Domain expertise
- Entrepreneurial
- Teamwork
- Results focused
- Communication
- Leadership

When the sales function is a separate unit from the marketing group, teamwork and communication has added significance because of inter-team coordination requirements. Selling is closing the deal, getting the customer to buy the product or service. Collaborative support from marketing is essential for sales to be successful. And sales, where it is more than order taking, is where the performance requirements are ultimately met or not. The two groups need to develop common purpose, make effective and efficient decisions jointly, and clearly define accountability for each of their units.

Team decisions about roles and responsibilities often become the issue when the market and sales effort are combined. The following case clarifies this point.

Organization: A 20 person advertising firm

- Products and Services
  design, development and placement of advertising
  consulting on advertising strategy for customers
- Markets
  chain retail stores
  small businesses
- Structure
  managed by the two owners
  a team of managers who both sold and developed the advertising. Group headed by a V.P. (not one of the owners)
- Primary Issues/Problems
  the team of advertising people needed to learn to sell, which was one of their responsibilities. Most of them had to sell a good deal to make the required corporate income.
  there were some major issues of the relationship between the V.P. and the other managers

Some persons in the group did not want to sell, or had little interest in developing their skills at selling, even though that was a responsibility all needed to accept because all of the account managers needed to be productive if they were to succeed financially. The way the group was structured combined selling and marketing, but accountability was unclear. As a whole, they were a very creative group, but in at least some cases they had not translated their creativity into effective problem solving/selling for customers. There was a good deal of resentment and interpersonal conflict over responsibilities for selling versus creating the advertising.

The Vice President had not been able to work through the internal team differences over roles and responsibilities and the fit of individual identities to selling, servicing, and creating the advertising. The group made progress at learning to sell, though progress varied a good deal with the individual. Some progress was made on the accountability for roles and responsibilities, but it was strained by distrust of the Vice President. The owners were distraught over how to fix the issue, and after some months, the Vice President left the organization and one of the owners retired.

The vague combination of the dual function of marketing and selling creates many of the team issues in many groups. That is complicated by the organizational structure dilemma. Put selling and marketing together and the problem becomes who does what. Keep the functions in separate units and the issue become intergroup collaboration. In either case, the basic solutions have to come from the team(s) themselves. They have to take accountability for building common purpose(s) and making decisions about how best to accomplish that purpose.

Those planning marketing, selling, and supportive service have additional challenges in today's high technology world. Technology produces situations where the marketing, selling, and servicing mix becomes even more complicated than usual. If the primary contact with the customers is through telemarketing, then those taking the orders may not understand that they are either seeking to close the sale, or to deliver service after the sale is complete. Sometimes telemarketers do both, at times creating functional confusion in their minds and in the minds of the customer. On the other hand, technology offers opportunities for marketing research because of the reach of the internet or related technical systems. Telemarketers can gather at least preliminary data on customer demographics and needs.

These opportunities sometimes lead to the tendency to conduct marketing, selling, and servicing without sufficient planning and attention to detail. When done poorly this can increase the cost to the customer base, creating problems for the telemarketing staff. The major point here is that technology has often confused the traditional functions of marketing, selling, and servicing. Organizational/team leadership are well advised to clarify the ways in which these functions are responsibly carried out.

Selling is identifying customer needs/wants and placing the features and benefits of the product/service next to those customers needs to build value.[8] That is true even in the impersonal world of selling through web sites. Servicing is maintaining or building the value of the product/service through delivery, clarifying product use, repair and return policy, and collaborative relationships with the customer.

The two are connected: selling depends on customer service to maintain or build value with the customer; good customer service builds the possibility of additional selling to repeat customers, or to people they refer. But selling and servicing are not the same. Customers benefit from knowing whether they are being sold or receiving service. Employees vary in their motivation, attitudes and abilities in selling or servicing, but not necessarily in doing both.

All work organizations do marketing, selling, and servicing to some degree. Government agencies decide who their clients are, provide information about what they do, and are usually interested in providing services so that their residents are content. For-profit corporations make similar decisions, whether it is a sophisticated marketing, selling and servicing structure such as that at Anheuser-Busch, or planning from with limited data by the CEO of a four person company. In any case, organizations need to be clear about who is accountable for which of the three basic functions and how they are to be done.

F. Production teams

Teamwork in production teams should focus on their basic function which is creating the product or service. Products are tangible, while services that we sell or provide are intangible. But both have to be created, modified, or acquired for sale and delivery to the customers or clients. The key is to meet the needs and wants of our defined markets. This is no different for most public or not-for-profit organizations than it is for the for-profit corporations. Increasingly, both for-profit corporations and public or community agencies are constrained by budgets, and are evaluated in terms of the outcomes they achieve for the money they spend. For example, cities are concerned with attracting people to their recreational centers and are competing with local golf courses, swimming pools, or other recreational offerings. Churches compete for congregations and provide those services they think meet their congregation's spiritual and temporal needs.

So production is important in virtually any organizational setting. People working together to enhance effective production is as basic as anything there is to organizational performance. A review of a case discussed in a different context in Chapters 4 and 7 will clarify this point and provide lessons for good teamwork in production.

Company: Tone's Brothers

- Products and Services
  processes spices

- Markets
    retail chains
    food manufacturers
    cafeteria management companies
- Structure
    plant of approximately 1000 people
    about 200 managers, sales, and customer service personnel
    unionized factory workers
- Primary Issues
    One of the company's plants in an eastern state was being closed.
    A production operation not previously used in Des Moines was
    being transferred from the east. The Des Moines plant needed to
    train its local employees in the new operation.

The Human Resources Department took responsibility to immediately
develop training for current and new employees in the operations of the
production machine to be transferred. A number of significant factors existed.

1. The product to be packaged by the new line was important to a
   number of Tone's customers. The purpose was to minimize interrup-
   tions in production.
2. The H.R. team had at least one trainer who had production experience
   and could with some effort begin to understand the engineering in-
   structions about how the machine operated. This trainer with the
   experience in production was willing to do what it took to develop
   the training.
3. The following areas of accountability were determined as a way of
   developing production training. The H.R. director would clarify policy on issues like making worker time available for classroom training
   on the operation. The trainer with operations experience would handle translating the engineering description of operations into every-
   day language. The training designer would focus on writing the training and training the trainers.

A team planning meeting occurred early, with decisions made on how to
proceed. The goal of developing the training and the associated deadlines
was clarified for everyone. Accountability for roles and responsibilities was
determined, and training decisions were made including the mix of classroom versus training on the actual machine. The project team began their
individual activities with all the pieces of the team plan in place.

After the training was designed, there was a "train the trainer" session.
One of the decisions had been to videotape the machine operations in the
eastern state before the plant was closed. The decision was made to connect
the training material to the videotape, adding to the reality of the explanations

of machine operations. The trainers being trained had both machine operations experience and some familiarity with conducting training.

While the pressures of the deadlines created some stress in the project, the team members worked well together. The actual training was rolled out on time, and production began very quickly. The training was evaluated by participants and received good ratings for clarity in explanations. The best evaluation of the benefits of this team effort, however, came when the newly trained operators working the newly established production line ran the company out of raw materials for the product, forcing a temporary shutdown. The good news was that the training worked extremely well and on time. The bad news was that the logistics team had not realized the impact of the training that was being conducted.

After a period of operations, human resources and the production team decided some additional development was needed. The machine had a number of trouble spots in the packing process, places where problems were most apt to occur. The production team needed to learn techniques for identifying and solving problems as a group to enhance output performance even more. The problem solving techniques were developed, taught, and used.

G. Conclusion

The use of teams and team-based structures in organizations varies a great deal with the philosophy of the leadership, the nature of the business, and unfortunately, whether teams are popular in the literature that year. Fads in performance improvement have included teams for customer service, team selling, quality circles, self-managed production teams, and project teams from matrix organizational structures. Despite the ups and downs in popularity, teams will always exist in organizations. The key is to make sure the organization is using teams at the right time for the right things (Teamwork and Decision Process). Finally, where teams exist, it is important to make sure they work well. (Model of Effective Teamwork)

H. Suggested action steps for organizational/team leaders

1. Identify your intact teams, those that are relatively permanent. Prepare a simple questionnaire using the three components of the Model for Effective Teamwork. Ask team members to evaluate their team performance on these areas. Make sure the survey asks for specific examples and illustrations, plus suggestions for improvement. Make this a confidential survey.
2. Take the results from the survey and begin to work on the areas of deficiency and suggested improvements.

3. Identify the marketing, selling and servicing functions in your organization. Are they complete, or is some function missing? Is accountability clearly identified?
4. If the functions are separated into different teams or units/departments, clarify what issues of interteam functioning may be occurring. Set up a process for finding common goals and effective/efficient interteam work.
5. Use the Teamwork and Decision Process model to see if teams are being used at the right time in your organization.

End notes

1. A reasonably good survey is in the early organizational behavior textbook by Hampton, David R., Summer, Charles E., and Webber, Ross A., *Organizational Behavior and the Practice of Management*, Scott, Foresman and Co., Dallas, 1982.
2. Aldag, Ramon J. and Stearns, Timothy M., *Management*, College Division, South-Western Publishing, Cincinnati, 1991, 560.
3. Schermerhorn, Jr., John R., Hunt, James G., and Osborn, Richard N., *Managing Organizational Behavior*, John Wiley & Sons, New York, 1991, 219.
4. The development of much of this model was influenced by a former colleague, Tony Montebello, and his recent book, *Work Teams That Work*, Best Sellers Publishing, Minneapolis, 1994, especially Chapters 3 through 8.
5. Cummings, Thomas G. and Worley, Christopher G., *Organizational Development and Change*, South-Western Publishing, Cincinnati, 1993, 217.
6. Pace, R. Wayne, and Faules, Don F., *Organizational Communication*, Prentice Hall, Englewood Cliffs, 1989, 210.
7. Internal document prepared by executives of a Fortune 500 company. Names withheld in accordance with company policy.
8. For an insightful discussion of the "communicating value" versus adding value in selling, see the following: Rackham, Neil and De Vincentis, John, *Rethinking the Sales Force*, McGraw-Hill, New York, 1999.

chapter nine

Leadership and teamwork

A. Introduction and linkage

The relationship between leadership of an organization/independent team and teamwork generally in the culture or within a functional team contains many directions.

1. The leader of the organization or team sets the tone for when and whether teamwork is used. While it is often true that the leaders often make the mistake of discouraging teamwork when it could increase quality and acceptance of decisions, the opposite is also true. Some leadership overemphasizes teams, so that the frequency of "meetings" and team involvement in areas where they do not belong end up interfering with performance. Leaders need to know when to use teamwork and when to not use it (discussed in Chapter 8).
2. The attitude team leaders have about the ability of their team to perform is probably the most critical factor in actual team performance.
3. Leaders need to make sure teams have a specific strategic vision or clear goals to accomplish the specific strategic vision if the team is dependent on strategic direction from the larger organization. Production teams are an example of groups who have their strategic direction handed to them. But even teams where the basic products/services they provide are determined by others need to have goals or standards for diagnosing current performance and targeting performance improvement interventions.
4. Leaders of organizations/teams directly affect performance of their teams in a number of ways: how they recognize and provide feedback on performance, the way they define their role and that of managers under them in terms of developing subordinates, and the way "mistakes" are handled. Two extremely different negative leadership attitudes about performance clarify the impact of leaders. In some cases, poor performance is expected and tolerated, and there is little belief in performance improvement or attempts to develop people. In another leadership direction, any performance that is not exactly at the

level expected by leadership is punished. In the first case, people are not motivated to get better at what they do. In the second case, people are afraid to be anything less than perfect as defined by the leadership. Learning from mistakes is not possible because mistakes are not tolerated.

Neither of these common leadership styles is productive of effective team effort.

B. *The importance of leadership in organizational/team performance: Introduction*

In recent years, leadership development has become a central theme in management literature and in performance improvement initiatives in organizations. In their 1998 report on training in organizations, ASTD listed 93% of all organizations offering courses on Management-Supervisory skills. These programs frequently offer suggestions and techniques for leadership for people at various levels of the organizational structure. In the same year, 63% of the organizations had Executive Development Programs, again undoubtedly having leadership components.

Further evidence of the popularity of the leadership topic is the enormous number of books available today on leadership vision and the changing role of managers. Some of these publications have been referenced in earlier chapters. But a good deal of confusion exists about the topic of leadership, despite its popularity. First, the term *leadership* and management tend to be used interchangeably, though it is useful to distinguish these two areas of knowledge and skills. In general, management is the performance of the long recognized management functions: planning, organizing, directing and evaluating. Staffing (selecting people to fill positions) and representing the team should be added to the list of management functions.

Leadership is the use of influence to encourage, direct and coordinate persons toward the accomplishment of the organizational/team goals. People who have the qualities or characteristics to exert this influence are called "leaders" or effective leaders.[2] In a sense, "ineffective leadership" is an oxymoron, though that phrase is used when people are talking about those having formal positions of leadership, such as a CEO, but not being good at the use of leadership influence.

As discussed in Part I, the first requirement of leaders is making sure that the thorough vision of the desired performance of the organization/team be defined and communicated. Then, leadership needs to provide the supporting systems and resources, including personnel with knowledge, skills and motivation, to get the work done for achieving the strategic vision and the supporting goals and standards. At this point, being good at the management functions is important in creating the performance necessary for achieving leadership's strategic vision.

Ideally, therefore, those heading organizations and teams should be effective at both management techniques and functions and at leadership. People good at managing but not leading have trouble getting people to follow their plans, organizing efforts, directions and evaluations. Those having great ability at developing and communicating a strategic vision, but little ability in, or patience for managing, get a lot of people inspired, but comparatively little actual performance occurs. In fact, many heads of organizations/teams are better at managing than leading, or vice versa. But the ability to lead and to manage is to a substantial degree learnable, as the following review of a model discussed earlier will demonstrate.

C. Sources of and changes in an individual's leadership/ management behavior

The Model for Growth previously discussed in Chapter 5 (Figure 5.1) and reprinted here as Figure 9.1, helps clarify the sources of a person's leadership/management style.

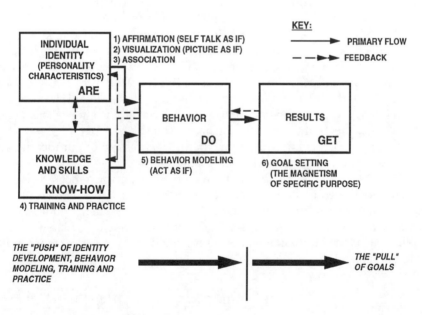

Figure 9.1 Model for Growth. (From Weaver, W. C., Achievement Associates, Inc., St. Louis, © 2000. With permission.)

Readers may remember from earlier discussions that the first box, "Identity," represents who each of us is as a person. This includes personality, motivations, attitudes and values, and other parts of the human being's identity. While some portion of our identity comes from heredity, how much is continuously debated, some part comes from conditioning or experiences we have had over life. That means that our identity is partly what we have done and where we have been. And, as has been said by many personal development experts, we are also right now "becoming" what we will be in the future. If our identity to some extent changes with our experiences, then it is logical to believe that when we change our experiences, we will, over time, also change our identity. This is true of any type of experiences we have, including those of leading and managing.

We also know from the Model for Growth that our behavior, which is influenced by our identity, is also influenced by our knowledge and skills. This includes our knowledge and skills in leadership/management, as well as the rest of life. That is the basis for the billions of dollars spent on leadership/management courses offered by organizations. The objective is to change the performance (behavior) of managers and leaders.

We all change over time; that is difficult to deny. The primary question is the extent to which people can manage their own change. "Growth" is desirable change over which the person changing has influence. The Model for Growth says that if you change your behavior through the techniques described here and more fully in Chapter 5 and the change in your behavior results in reaching your goals (improved performance), then the change will become permanent because the individual likes the improved results.

Our identity, especially attitudes and personality and our knowledge and skills drive our behavior including our behavior as managers/leaders. But so does the behavior of others who have power or influence over us. This is "Behavior Modeling" as described in the Model for Growth. Most managers and leaders who are not at the top of the organization, report feeling that "what is accepted management style" in the culture, or by their own boss, affects their behavior. That is called "role playing," which is managing our behavior away from a natural expression of our identity and knowledge and skills toward what is expected.

If "role playing" leads to more effective performance as a manager/leader, then it is desirable, even though it is usually very uncomfortable. The key questions, long a thorny issue in management and leadership discussions, is whether there is a model of management/leadership worth following as the individual works at growing their own performance in this vital organizational role.

D. Importance of leadership and management in organizational/team performance revisited

One way to understand the importance of management/leadership in the performance of organizations and teams is to refer to the model for Organizational Success in Chapter 3 (Figure 3.1) and reprinted here as Figure 9.2.

For a brief review, remember that the Strategy is answers to questions about the following: What products/services do we provide, what markets

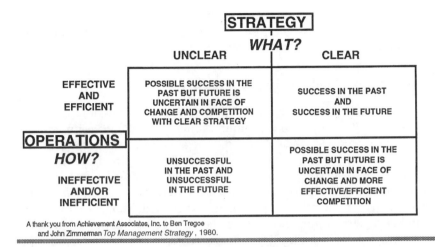

A thank you from Achievement Associates, Inc. to Ben Tregoe
and John Zimmerman *Top Management Strategy* , 1980.

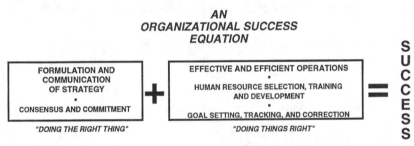

Figure 9.2 Model For Organizational Success. (From Weaver, W. C., Achievement Associates, Inc., St. Louis, © 2000. With permission.)

do we serve, what methods of marketing and distribution do we use? This is where leadership must first perform effectively if the organization is to be successful. The need is to develop vision and communicate it to those in the organization/team.

Operations are answers to questions such as: How do we hire? How do we manage, recognize and reward? The definition of how these things should be done is the responsibility of leadership. The specific initiatives, programs, and daily support to make sure the operations are effective and efficient is largely the responsibility of management.

Somehow, therefore, the establishment of a clear strategic vision and the management of daily operations has to both be done for an organization to be successful. While this is hard to know for sure, it is probably true that most organization/team managers are generally better at managing than they are at leading. In the case of a leader who has trouble with the strategic vision requirements of leadership, using a top team to develop that vision is even more crucial to organizational success. Those who are good at building and communicating a strategic vision, but less capable at daily management, need a team of managers who help them with daily operations. Occasionally, a person becomes effective at both leadership and management and has a major role in both strategy and operations.

E. Model of effective management/leadership performance

What style of leadership/management is most effective at accomplishing the multiple roles of defining and communicating the strategic vision, making sure that the systems are there to make strategy happen, and helping ensure that daily operations are effective and efficient? For much of the history of the writings in western civilization, this has been a central concern and area of disagreement. Plato gave us *The Republic*. Machiavelli wrote *The Prince*, which is where the word "Machiavellian" originated. Many authors in the 19th and early 20th centuries discussed the "captains of industry." In more recent times, we got contingency management and "situation leadership," the latter which was succinctly described in the following quote.

> According to Situational Leadership, there is no one "best" way to go about influencing people. Which leadership style a person should use with individuals or groups depends on the readiness level of the people the leader is attempting to influence.[3]

Situational leadership rejected what has been called "normative approaches," the belief that there is a best way to manage and lead. After centuries of disagreement over the best leadership approach, moving to approaches than were more "adaptive" to people and situations was a natural occurrence. In the 1970s, however, a series of studies began that led to a data-based model of "the best" management/leadership style. The Achieving Manager Project, based on studies of achievement with thousands of managers over three decades, resulted in a model of effective leadership/management, The High Achieving Manager Model.[4]

There were two primary issues to be answered in this gigantic study of managers. First, once and for all, can we find some way to define and measure leadership/management effectiveness so that one group of leaders can be identified as more effective than other groups? That is, can we identify successful leadership/management performance in an applied way so that we can measure it in the real world of organizations?

The second issue has to do with figuring out what makes successful leaders/managers more effective than others. If we can find a group of High Achieving Managers, what do they do to be high achievers compared to other leaders/managers?

Defining high achievers in organizations was tricky. What was needed was a measure of achievement which was objective in the sense that all could look at it and at least generally agree on the results of the measurement when it was done with actual managers. In addition, the project director, Dr. Hall, believed it was necessary to use measures of achievement that were important to actual managers. They decided to use the Managerial Achievement Quotient for distinguishing the performance of managers. That is, they developed individual scores on achievement for each manager in the study by giving them points for their success in the organization.

Here is how it worked. The original study in the 1970s was done over about 50 organizations with over 12,000 managers. The achievement of each manager was rated based on a formula measuring their promotion rate compared to their age. Since promotion usually involves increase in salary, responsibility and number of reports, promotions also indirectly measure other achievement factors, especially since the date covered thousands of managers from many different types of organizations. Using success in promotion rate compared to age, the results are illustrated in Figure 9.3.

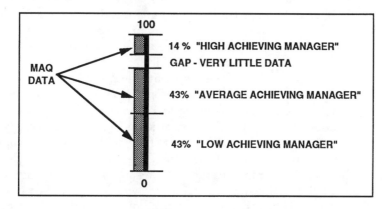

Figure 9.3 MAQ Illustration.

The rating scale on promotion rate was "normalized" mathematically so that the top scoring manager's score was adjusted to be 100 points, and the lowest manager's score was adjusted to be equal to 0. Then all other managers were placed on the above diagram in numerical relationships to the range from 0 to 100. The results were dramatic. In the 1970s version of the study, the High Achieving Managers, those being promoted more rapidly than others, equaled about 14%. There was a gap from there to the cluster of managers scoring in second place. So some 1680 managers (14% of 12,000) were high achieving as defined by promotion rate and the related growth in power and rewards.

When discussing this part of the Achieving Manager Study, organizational/team managers are usually impressed with the magnitude of the sample group and the potential benefit of finding a model of leadership based on what thousands of managers do. Some reservations about the methodology do, however, occur. One is that promotion rate may not be a direct measure of achievement. Promotions are sometimes political, or influenced by nepotism, sexism and racism, among other factors. Aren't those factors involved in the success of the High Achieving Managers? That argument would be more convincing if this were a very small sample of managers, instead of 12,000. It is reasonable to assume that the size of the sample of managers and the fact that they are from many different organizations means that many elements were involved in their success. This large group of managers was clearly more successful in promotion rate than others, and that success was probably related to both performance and non-performance factors.

A subtle extension of the above concern about the model sometimes mentioned by organizational/team decision makers is concern with why the research group did not measure performance directly. They ask the following: "Even if the MAQ does measure performance achievement by measuring promotion rate, why not measure goal achievement or use the performance evaluation system in the organizations?" The problem is that many organizations have only fuzzy or ill defined goals, so how can we know if they have been achieved? In addition, many organizations either do not have performance evaluation systems or they are vague in what they measure, and are themselves very subjective and political.

If you are willing to accept the notion that over a large sample of managers from many different organizations, promotion rate in relationship to age, is a measure of performance, the model becomes a powerful one. Two basic considerations are usually the greatest factors in accepting the Achieving Manager project as valid. First, the same methodology was used again in the 1980s with thousands of different managers and different organizations from the 1970s version of the study. The results were largely the same, a small group of High Achieving Managers who were being promoted more rapidly than others. Finally, on this point, the study was replicated one more time in the 1990s, again with a large number of managers from different organizations, and the results of this third version of the study were repeated again. In these three decades, there was a percentage of thousands of mangers who were able to do whatever it took to achieve promotion rates more rapidly than about 86% of all the other managers in the sample.

The second reason why most organizational/team leaders introduced to the High Achieving Manager project accept the validity of the model is this. When the study measured how the total sample of managers worked with their direct reports, the results were dramatically different for High Achieving Managers compared to low and average achievers. This difference in how high achievers managed and lead occurred occurred in the 1970s study and again in the replications of this project in the 1980s and the 1990s. The High Achieving Managers behave differently than other managers when working with their direct reports. While there is no direct evidence that their style of leading/managing itself led to their promotion rate in the organization, it clearly did not stop that success at climbing the organizational ladder.

Let's return to the Model for Growth (originally introduced in Chapter 5, now Figure 9.4) to review why the validity of the Achieving Manager Project is so important.

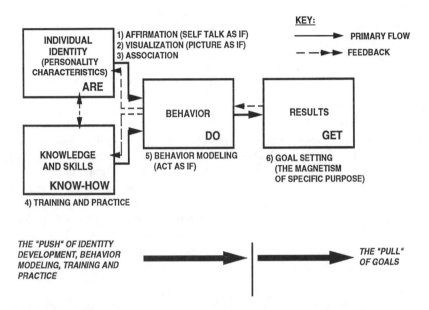

Figure 9.4 The Model For Growth. (From Weaver, W. C., Achievement Associates, Inc., St. Louis, © 2000. With permission.)

One method by which people can grow their performance in any area is to find a model in that area that they believe is valid and effective. We are all influenced by models throughout our lives, but especially when we are inexperienced in some area and have not developed a style. So, new employees are especially susceptible to "learning" management style from those they see in their new organization. They will then accept one model, or elements from a number of leadership/management models, which they will come to use when they themselves become managers. Why people accept one model over others is complicated, but undoubtedly has to do with values, attitudes, personality factors, and how much of one managerial/leadership approach they see compared to others (conditioning).

Therefore, accepting a model of management/leadership as valid means leaders/managers can choose to try and become more like the manager they believe in most. They can use that model to directly change their behavior. The key to becoming the Achieving Manager, is to learn to act like one.

So what did these High Achieving Managers do differently than the Low and Average Achieving Managers? To answer this major issue, the research team look at six areas of management and leadership. They were as follows:

- Management Values (Belief about people generally and workers specifically)
- Involvement, Access, and Participation (Now called Participative Management)
- Communication Style
- How Managers Seek to Motivate Others
- Power and Empower (How They are used)
- Overall Leadership Style

In the Achieving Manager Projects, the research team adopted models in each of these six areas of management/leadership and used validated surveys to get feedback from three direct reports of each of the 12,000 managers. So each manager was described by three direct reports in terms of how they behaved in each of the six areas listed above. But they were also being evaluated, even thought the direct reports did not know they were doing that. The evaluation came from the fact that each of the High Achieving Managers tended to score much like other High Achieving Managers in each of the six areas of performance listed above and differently from the 86% of low achievers and average achievers. So the scores of the High Achieving Managers become a desired result and is a measure of performance leading to high achievement in the leadership/management role. What scores do High Achieving Managers get from their direct reports in each of these six areas?

F. Management values (beliefs about people and workers)

In measuring manager values about people, the High Achieving Manager project used a values and attitudes model (illustrated Figure 9.5) from a classic work by Douglas McGregor, *The Human Side of Enterprise.*

Direct reports from the Achieving Manager Project completed instruments that measured how they were treated by their managers. They were *not* directly evaluating their managers on X and Y, but the survey instruments led to scores on X and Y for the managers described and identified how much X versus Y they exhibited.

Figure 9.6 shows how the X and Y results were significantly different for the High, Average, and Low Achieving Managers.

The implications for these results comes from the "Pygmalion Effect," the self-fulfilling prophesy. This is the idea that we get what we look for and what we expect from others, including those who work for us. Keep in mind that one of the most significant factors in the performance of teams is the expectations about the team held by the leader. That is also true of expectations about individual performance. The manager with a lot of Theory X in their beliefs about workers expect people on the whole to dislike and want to avoid work and to need to be controlled if any performance is to occur. The Theory Y manager sees people as believing work is natural, and having wide spread potential for growth. Perhaps the most significant factor is that the High

DOUGLAS McGREGOR'S THEORY "X" AND THEORY "Y"

"X" BELIEFS SEE PEOPLE AS:	*"Y" BELIEFS SEE PEOPLE AS:*
* DISLIKING AND AVOIDING WORK	* SEEING WORK AS A NATURAL AND POTENTIALLY FULFILLING PART OF LIFE
* NEEDING TO BE CONTROLLED, AND EVEN COERCED, IF THEY ARE TO PERFORM	* CAPABLE, UNDER THE RIGHT CONDITIONS, OF SELF DIRECTION AND SELF MANAGEMENT
* AVOIDING RESPONSIBILITY	* ACCEPTING RESPONSIBILITY AND UNDER THE RIGHT CONDITIONS, SEEKING ADDITIONAL RESPONSIBILITY.
* SEEING THEIR PERSONAL GOALS AS INHERENTLY IN CONFLICT WORTH THE GOALS OF THE ORGANIZATION	* RECOGNIZING THAT CONTRIBUTING TO ACHIEVING THE ORGANIZATION'S GOALS CAN ALSO CONTRIBUTE TO ACHIEVING THEIR PERSONAL GOALS
* NOT TYPICALLY CREATIVE	* HAVING WIDESPREAD POTENTIAL FOR CREATIVITY
* HAVING LIMITED GROWTH	* AS HAVING WIDESPREAD POTENTIAL FOR GROWTH

Figure 9.5 Douglas McGregor's Theory X and Theory Y.

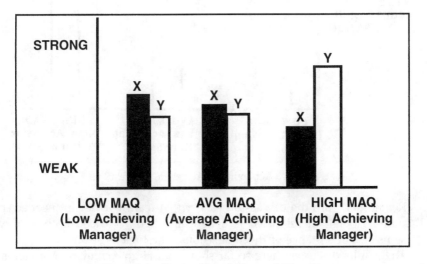

Figure 9.6 Theory X and Theory Y Results.

Achieving Manager believes people recognize that contributing to help the organization achieve its goals can also help them achieve their own individual goals. Motivation comes primarily from the individual's expectation that if they help the organization/team achieve they will get their needs met.

Using Theory X does not usually increase promotion opportunities for managers, though those in fear oriented organizational cultures seem to act as if it does. It is important to note that the High Achieving Manager is not a "non-manager." They expect and work to achieve results. They provide support, clear direction, and feedback on performance and opportunities for learning. But they expect performance.

G. Participation, access, and involvement (participative management)

The Achieving Manager Project also generated feedback for managers on uses of participative management techniques (Figure 9.7). In participative management "the emphasis is on joint decision-making about events which have future implications for the parties involved, and over which they can realistically exert influence."[4] It turned out that High Achieving Managers made more use of participative decision-making as measured by feedback from their direct reports.

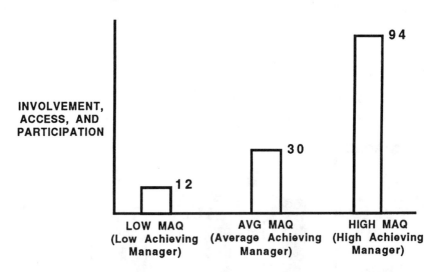

Figure 9.7 Achieving Manager Project results.

Remember from the discussion earlier that teamwork is not needed in all settings for all decisions. It is beneficial when you need both quality decisions and acceptance of those decisions. The logic is strong.

The results discussed here so far show that High Achieving Managers, those more successful in their organizations measured by promotion rate and related increases in power and rewards, expect performance from workers and believe they can perform. The High Achieving Manager also uses participation as an important part of their style, especially when they are looking for quality decisions and acceptance of those decisions. Management values and participative management are connected. If a manager has high expectations for the performance of their workers, they are more apt to use participation to get the valued input from their people. If they have negative expectations, why have the workers involved in deciding anything?

H. Interpersonal competence and communication style

Communication is one of the areas seen by many organizational/team leaders as the greatest source of performance problems. The Achieving Manager Project staff had the insight to include communication as part of the assessment of managers. The model they used, the Johari Window (Figure 9.8), remains one of the best Models of Effective Communication.

The Johari Window focuses on the importance of two dimensions in communication: asking for feedback and sharing information (exposure). When a person does both, they have an *arena*, "where the action is." When they do not ask for feedback, they are operating without information, so they have a *blind spot*. When they do not share information and feelings, they are keeping things to themselves, so that is their *facade*. There are four basic styles of communication resulting from how good a person is in asking for feedback or sharing information. Those styles are presented in Figure 9.9.

The type A style is very closed, with little effort at getting feedback and little sharing of information. This style of communication is troublesome in any situation requiring coordination or teamwork between people. Type B style is all asking for information or feedback and does little sharing. The popular television detective Columbo represents this style extremely well. He kept asking questions until he got his criminal. The problem with type B style of communication is that people feel "interrogated" when the person with this style only asks questions and does not share their information. Fear, especially if the type B person has positional power, is natural and shut down in the conversation usually occurs.

The type C style is frequent among top level executives, especially in the business world. This type tells people more then they ever needed to know, but fails at feedback reception. They do not listen much, but they tell a lot. An excessive and on-going use of this style is often a reflection of arrogance and over use of positional power.

Type D is someone who both listens to others and shares information as needed. That style is critical if teamwork or collaboration is going to occur. But most people have a tendency to more often use types A, B, or C. The reasons for our communication style are multiple, but they include the following: learned habits from childhood, personality, and the culture of the organization in which a person is working.[6]

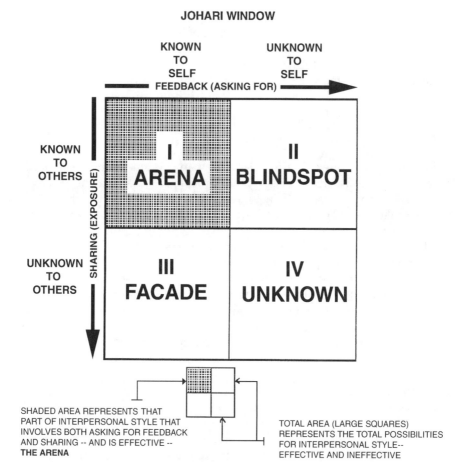

Figure 9.8 Johari Window.

By looking at Figure 9.9, readers can see that the High Achieving Manager is type D in the Johari Window assessment by direct reports. Again, the direct reports simply describe behavior, and the results of those description lead to a Johari Window score. The direct reports did not "evaluate" their managers by comparing them to the Johari Window. But their description of the manager led to the score. The High Achieving Manager was half again better than the Average Achieving Manager in both feedback and sharing information. The Average Achieving Managers tended to score like the average person in the society in communication effectiveness. The Low Achieving managers were the type A closed communication style.

JOHARI WINDOW COMMUNICATION STYLES

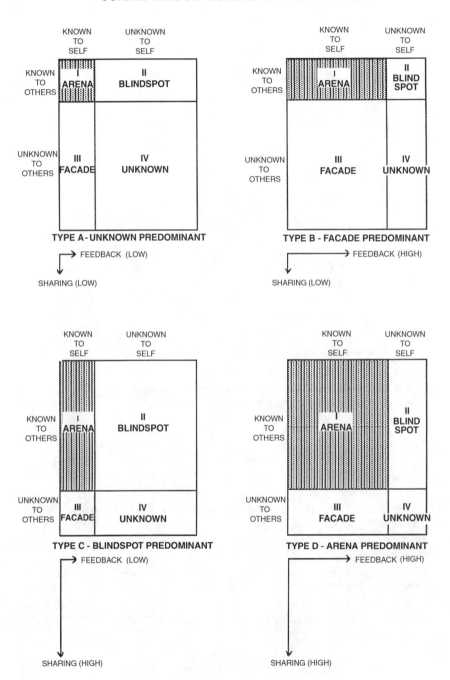

Figure 9.9 Johari Window communication styles.

So, the High Achieving Manager has respect for people's ability to perform and expects it. (Theory Y) He or she uses participative management where appropriate and communicates openly with direct reports.

I. Motivation: The manager and those he manages

Another topic many organizational/team decision makers believe is central to performance success is motivation. As is true with the other areas of management/leadership, having a model for focus of discussion improves clarity and understanding. The classic works of Maslow and Herzberg help clarify the Achieving Manager results (Figure 9.10).

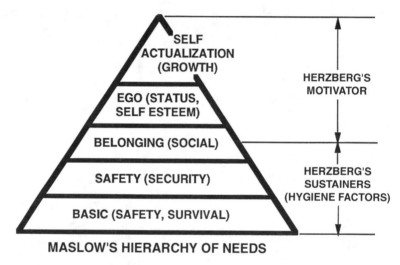

Figure 9.10 Motivation Illustration

Feedback from the direct reports in the Achieving Manager project indicated that Low Achieving Managers depended heavily on Safety and Belonging, both with themselves and in managing others. Average Achieving Managers focus on Ego and Safety, the former motivation is consistent with the image of a "hard driving, aggressive and ambitious" executive. The High Achieving Manager, while paying some attention to things like salary, benefits, and equipment, focuses heavily on team work (Belonging), recognition (Ego), and developing people (Self-Actualization). Again, the consistency of the results is remarkable. High Achieving Managers expect their people to perform, use participative management, communicate openly with their direct reports, and use "higher level" forms of motivation with themselves and with others.

J. Power and empowerment

Managers use power repeatedly. Power is influence in making decisions or the ability to make something happen. The use of power is central to leadership/management and how it is used goes a long way to defining the style of a leader/manager. Power use starts with hiring decisions, continues with defining roles and responsibilities, and involves providing evaluation on the performance of subordinates. But power reflected in decision making is also present in defining the strategic direction, setting strategic objectives, and managing the group's performance in achieving those objectives. Setting up a system of cascading goals throughout the organization and deciding what standards or Key Performance Indicators need to be measured is also an important use of power. As noted earlier, leaders tend to get back from others what is being measured as important.

There are two parts to the model for assessing power use in the Achieving Manager Project: Power Motivation and Power Sharing. The Power Motivation model originates from the famous team of McClelland and Burnam, the former a psychologist who contributed a great deal to our understanding of the achievement drive. Basically, there are three types of power motivation (shown below).

Personalized Power	Socialized Power	Affiliative Power
* Self-serving	* Team oriented	* Concerned about being liked
* Often arrogant	* Goals of the group	* Trouble with confrontation

Low Achieving Managers were seen by their direct reports as having more Affiliative power than either Personalized or Socialized power. Average Achieving Managers were focused more on Personalized power. While all three categories of managers had each of the power motivations to some degree, the High Achieving Manager's greatest source of power motivation is Socialized power.

Power sharing is the second element in the power model. In summary, Low Achieving Managers tend to give power away. Average Achieving Mangers keep it for themselves. High Achieving Managers share power with others, which is what collaboration is about. This is not surprising considering their Theory Y values and attitudes about their direct reports and their practice of participative management. Remember, however, that the High Achieving Manager is only about 14% of the total group of managers.

K. Leadership style

The sixth area of leadership/management and the one with the most impact studied in the Achieving Manager Project was leadership style. The model used to understand leadership style was the Managerial Grid (Figure 9.11), developed by Mouton and Blake and still remarkably popular today some 35 years after its creation.

This model uses two elements of leadership style long discussed in leadership literature, training programs, and discussions. The two elements are concern for getting the job done, and concern for people. Listed below

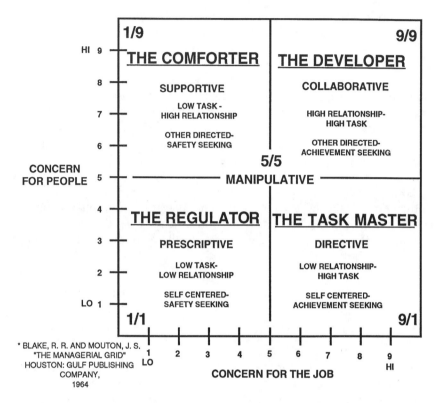

Figure 9.11 Mouton-Blake leadership styles. (From Blake, R. R. and Mouton, J. S., *The Managerial Grid*, Gulf Publishing, Houston, 1964. With permission.)

are five basic leadership styles resulting from combining the amount of concern held by leaders/managers regarding the two elements.

- **9/1 Task Master** — Powerful concern for the job, little for the people.
- **1/9 Comforter** — Powerful concern for the people and their comfort, little for the job.
- **1/1 Regulator** — Little concern for either the job being done or the people involved.
- **5/5 Manipulative** — Tries to balance concern for both people and job, not too much of either.
- **9/9 Developer** — Powerful concern for both the job and the people.

The results of the Achieving Manager Studies in all three decades are consistent. The High Achieving Managers had positive expectations about their people, used participative management, communicated openly with people, motivated toward higher level needs, and used power mostly for performing to reach the goals of the organization/team. It is not surprising, then, that the High Achieving Manager also tended to perform as a Developer more than the other four styles in the Leadership Grid. The small percentage of High Achieving Managers out of all managers consistently depended on methods aimed at expecting and getting the best out of their people.

The High Achieving Manager can serve as a very influential model for developing leaders/managers. Experience with a large number of current leaders/managers shows that this influence comes from three sources. First, many people responsible for leading and managing are "in search of a style" they can depend on. Every day there are new books and new leadership programs to attend. The questions many managers ask is which style should they believe in and use? What will make them most effective at leading organization/teams? A brief case will dramatize this point.

Organization: Ranken Technical College

- Products/Services
    technical education
- Markets
    adults
- Structure
    central leadership team of about eight people
    president and vice president with strong personalities
    two faculty groups: general studies and technical studies
    staff
    about 2000 students
- Primary Issues/Problems
    evidence from an organizational culture survey and discussions with some influential leaders in the organization indicated that power was overly centralized with too few decisions made by people who had to implement them
    some interpersonal conflict between the two top leaders and others in the organization such that performance was being damaged, partly because of the time and focus spent on unproductive disputes.

The leadership group of the college decided to go through a development process in which they learned the High Achieving Manager model. As part of this, they received feedback from at least 3 direct reports using survey instruments that duplicated the original Achieving Manager project. They then set goals and action steps to improve their style to become more like the High Achieving Manager. This included a series of meetings with the direct report to discuss specific steps in leadership/management improvement. The project lasted about six months.

When the group discussed the Participative Management/Power and Empowerment issues, they came to terms with the centralized decision making problem and decided to push many decisions further down in the organization. The combination of the improved style of the individual leaders/managers and the decentralization of decision making has resulted in dramatic improvements in their organizational culture. The college is very successful and is regarded as one of the three best technical colleges in the country. Now their organizational culture and decision making processes are also improved.

L. Suggested action steps for organizational leaders/managers

1. Review the status of leadership/management knowledge/skills and style in your organization. Do not depend on just a few opinions, use as broad a source of data as practical.
2. Identify areas where leadership/management is strong according to those being managed and where it is deficient.
3. Use the top group of leaders/managers to begin deciding on actions to improve leadership and management. Discuss what impact current deficiencies in this area are having on performance and how various efforts to improve style might also improve measurable performance.
4. Discuss the various options for improving leadership/management performance with those being managed and led. Recognize that the ultimate test of effectiveness of leaders/managers is the thoughtful opinions of those they lead.

End notes

1. Bassi, Laurie J., and Van Buren, Mark E., *State of the Industry Report*, Training and Development, Alexandria, January, 1998.
2. Aldag, Ramon J. and Stearns, Timothy M., *Management*, College Division, South-Western Publishing, Cincinnati, 1991, 500.
3. Hersey, Paul and Blanchard, Kenneth H. and Natemeyer, Walter E., *Power Perception Profile, A Summary*, Leadership Studies, Inc., San Diego, 1988.
4. For a discussion of the first wave of this study of managers, see the following: Hall, Jay, Ph.D., *Models for Management, The Structure of Competence*, Woodstead Press, The Woodlands, 1998.
5. Hall, Jay, *Models for Management, The Structure of Competence*, Woodstead Press, The Woodlands, 1998, 509.
6. Pace, R. Wayne and Faules, Don F., *Organizational Communication*, Prentice Hall, Englewood Cliffs, 1989, 121.

chapter ten

Performance improvement and developing the individual

So far we have discussed performance improvement at the macro and team levels. Interventions discussed include strategic planning and cascading goals, building a learning organization, and team development. What is the role of the individual employee's performance in all of this?

1. Individuals create the strategy, set the objectives and either do or do not make it happen. The CEO and individuals on the leadership team can support and recognize performance and deal with reducing performance gaps. Individuals, including those in leadership, make personal decisions about "buy in" to the plan, and then deal with their attitudes and motivation about making the strategy work.
2. While they are influenced by group goals, roles, and cohesion, the individuals in marketing, sales and service and production react to the "teamness" in their own unique ways, for better or worse.
3. Perhaps nowhere is the power of individuality so apparent as in organization/team management. The Achieving Manager Project, for example, discovered that among 12,000 managers there were varying degrees of leadership success. High Achieving Managers dealt with their direct reports much differently from other managers (about 86% of all managers). In fact, experience shows that individual managers vary a great deal in how they use the six leadership areas identified in the Achieving Manager project (see Chapter 9).

> Any individual's performance is an interplay between organizational characteristics, team dynamics, and their identity, behavior, and goals and motivation.

Do organizations and teams have influence over the performance of individual employees? Despite the fact that individuals ultimately decide about their own motivation and desire to perform, organizations and teams do influence that decision in many ways. They select certain individuals and not others, they direct people more or less effectively, and they recognize and sometimes punish individuals. In short, they provide opportunities for learning and performance growth, or they provide conditions that result in performance failure.

A simple model, Identity/Role theory (Figure 10.1), helps clarify this as discussed earlier.

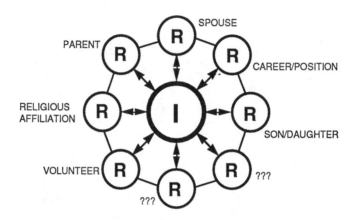

Figure 10.1   Identity/Role Theory

Identity is who we are, the same individual personality, motivation, attitudes and values characteristics discussed under Identity in the Model for Growth. Two primary forces have an impact on the way we perform. One is our Identity, and the other is what we know, know how to do, and believe about the role we are performing. Some of our work roles are heavily influenced by the organizations in which they occur (e.g., supervisory roles in a manufacturing company). Our personal roles are defined more by the culture and socio-economic levels in which our personal lives exist.

All of our roles in life are connected in the sense that how we are doing in one role will probably affect how we perform in others. This is very true regarding the connection between personal and professional roles. Few people succeed for very long at separating their personal and occupational lives. When things are going well personally, it is usually easier to perform well at work, even if work is not everything we would like. If things are going well at work, our personal lives benefit as well. The following simple formula helps communicate this.

Fulfilling life = Good Love + Good Work

As one CEO was heard to say, "When big problems occur in my company, I go home and kick the dog." The complicated interaction between the many roles in life are illustrated by Identity/Role Theory.

While the importance of personal versus occupational life varies with the individual, there is little doubt they are interconnected. Even examples of people who have little personal life and who try to lose themselves in work support this conclusion. They are motivated to substitute work for lack of personal satisfaction. The fact that employers recognize the personal life and professional life connection is demonstrated by the popularity of Employee Assistance Programs and the national movement to provide educational programs in personal finances. To continue this point, problems in personal finances are among the major causes of difficulties at work.

A. What do we know about developing individual performance?

Two conclusions about developing individual performance flow from the above discussions.

1. Any performance improvement program should make efforts at developing the individuals involved, whether as team members, employees of organizations, or single individuals needing special attention. Two examples will help clarify the point. When restructuring occurs in an organization, attention should be paid to how the newly connected individuals will interact. Personality profiling and what is known about previous work history of the involved individuals is useful here. Redrawing organization charts puts individuals together and can often result in conflicts that slow performance. That can and should be discussed and managed in the beginning of the reorganization.

   When providing training in leadership, selling, or customer service, for example, feedback to individuals on their identity and how that will have an impact on their behavior in these roles and it provides opportunities for improving performance. If a customer service provider discovers that they have a high degree of assertiveness, they can focus on developing listening skills so as to avoid initiating conflicts with demanding customers. More about this in Chapters 11 and 12 which concentrate on training.

2. Each individual makes a conscious/sub-conscious mix of decisions about how hard he or she will work to perform well or to improve his or her performance if that is needed. This seems to have something to do with how the individual sees the effectiveness of his or her own efforts at reaching goals, his or her own goals and those of the organization or team. This is called Expectancy and the well-known concepts used to describe its influence on individual performance is Expectancy Theory. Robert Smither states in *The Psychology of Work and Human Performance*, "Expectancy relates to the probability of an outcome occurring."[1]

3. The best way to have a positive impact on motivation is by finding the connection between the personal goals of the individual and the

goals of the organization/team. All individuals have goals, even if they are poorly defined. Goals come from our social and physical needs. Because the nature and strength of needs vary between individuals, goals which we seek to satisfy our needs also vary. So the best way to affect motivation is to identify the individual's needs, and establish an environment in which achievement of the organization/teams' goals give the individual what they want. The illustration in Figure 10.2 shows this clearly.

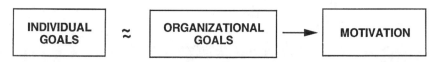

Figure 10.2   Model of Motivation.

Encouraging motivation among employees is one of the most common concerns identified by leaders of organizations/teams. But they tend to look for a single magic answer which will work for all individuals. The best approach is to identify the *different motivations* for each employee, and then seek to provide goals, rewards, and recognition that meet those motivations. That is how the High Achieving Managers, the small percentage of successful managers discussed in the last chapter, handle motivation.

Identifying motivations of employees is not as complicated as it sounds. People will identify their own motivations when asked. In general conversation, employees discuss the things that are important to them and what they want in life or at work. The biggest challenge for managers is to find ways to meet the motivations of others. Some people want tangible rewards, so salary and promotions are a key. Others want recognition for their work, so providing sincere, specific and timely feedback on performance is important to their motivation. Still others want challenging assignments, or variety in their work, while their co-workers want stability and predictability in assignments. The challenge is to meet the individuality of the employee.

B. The model for growth and developing individual performance

As discussed in Chapter 5, the Model for Growth (Figure 5.1) clarifies methods for developing individual performance in a number of ways. If a sales team is taught how to maintain positive attitudes through conscious use of affirmation and visualization, (we all do without knowing it), it may well enhance Identity and then improve sales performance (Behavior). If a team of managers decides to improve team problem solving by getting everyone involved in a seminar to learn teamwork skills, and then follows up by setting weekly team meetings, the managers may have increased their knowledge and their skills. They will have also changed behavior as they reschedule

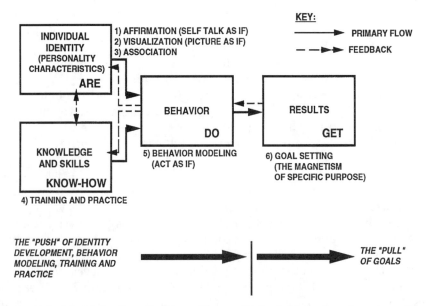

Figure 10.3 Model for Growth. (From Weaver, W. C., Achievement Associates, Inc., St. Louis, © 2000. With permission.)

work loads to make time for the team meetings. Finally, if they make progress from the team meetings on identifying and solving real issues that block goal achievement, they thereby improve performance.

The Model for Growth (presented again as Figure 10.3) is most fundamentally a set of six techniques for the individual to take responsibility for his or her own performance improvement. Working backward133 on the model, developing one's self has to do with setting clear goals and finding ways to adjust behavior to meet these goals. It also has to do with individuals finding acceptable models of effective performance to imitate as they stretch to reach their performance goals. A model of an effective salesperson, and the High Achieving Manager are only two examples. Knowledge and skills in the Model for Growth reminds us that learning is a life long commitment as the nature of technology, our jobs, and our organizations/teams change over time. Finally, the Model for Growth is a reminder of the importance of knowing one's self and how to get the most out of who one is while working on who one wants to become.

To some, the Model for Growth is suspect because it assumes that individuals can improve their own performance and that they want to do so. Developmental psychology provides some insight into whether we can develop ourselves, albeit with organizational/team support where possible. One recent excellent book on developmental psychology by Carol Sigelman and David Shaffer provides the following summary of basic issues regarding developing the individual.[2]

1. The question of nature versus nurture has long been a focus of debate. For example, are leaders born or are they developed? "On the nature side of the debate have been those who emphasize the influence of heredity, universal maturational processes guided by the genes, and biologically based predispositions."[2] However, "On the nurture side of the debate have been those who emphasize environment — forces outside of the person, including life experiences, changes achieved through learning, and the influence of methods of child rearing, societal changes, and culture on development."[2]

Sigelman and Shaffer argue that most developmentalists today see development as an "interplay" between both nature and nurture.

2. The second question regarding developing the individual is whether we as individuals can take an active role in our own development by planning our learning and life experiences. Or are we destined to be passive recipients of the influence of forces beyond our control? This basic question is a complex one. Like other questions of great complexity, there may be no final answer. Theorists disagree about how active individuals are in creating their own environments and, in the process, producing their own development.

Since the experts do not agree on "the activity/passivity issue," each of us can decide whether to sit back and let the forces have their way with us or whether we will make efforts at influencing our own world so as to develop in ways we prefer. This is not a question reserved for philosophic debate. For example, many managers seeking help in improving their leadership/management style wrestle with the question of whether they will take responsibility for improving their performance in the role of leaders/managers. Some decide to make the effort. Others decide they cannot learn better ways or it's not worth the effort, or the organizational/team culture is too powerful an influence on them.

Sigelman and Shaffer's book referenced above, and others in the field, argue that development is a life long process, not one occurring only in the early years of life in contrast to the view maintained by psychology a few years ago. Whatever the subtleties of how we develop, and whether we can consciously influence our development, it is clear that in today's world one is never too old to be a developing person.

C. *Morale and individual performance*

Morale is the set of attitudes employees have about their working situation which has an impact on performance in a number of ways. Developing the individual sometimes involves assessing the current status of morale and identifying and acting on morale where it is deficient. Morale tends to be the result of attitudes about three areas of the work situation listed below.

1. *Attitudes about the organization* includes as number of areas of common concern to employees. These usually are at least the following: how the organization treats its people and whether the organization does

good work and provides a desirable and quality group of products and services. Attitudes about the organization also include how well the organization is managed.

2. *Attitudes about one's boss* is a major source of morale, good or bad, for most employees. This can range from adoration of the boss to the more frequent contempt for the boss because of poor treatment. Questions or topics of discussions about bosses tend to include complaints that the boss is too demanding and has little tolerance for mistakes, or the boss is too easy; it's like not having a boss at all. Others include the boss treats people fairly, or has favorites; the boss provides support in job performance, or totally ignores employees and what they do.

3. *Attitudes about one's job* are also a source of morale that is often overlooked by organizations seeking to understand this organizational component. Issues include the following: the job is fun and enjoyable, or going to work is a pain; the employee feels competent at what he does, or realize there is a lot he is not good at.

Actions for improving morale begin with knowing in what areas morale is an issue. As examples:

- Establishing and/or communicating a clear strategy can significantly improve morale about the organization and how well it is managed.
- Establishing a performance management system ("appraisal system") with an employee development component can help morale in all three areas identified above.
- Working to improve leadership style and management skills in the organization/team goes a long way towards improving morale about one's boss, as long as the bosses are trying to improve in a sincere way. Conscientious efforts at improvement by managers gains a lot of credit with employees who see their bosses working to improve style. Actually improving leadership/management style is also received well by employees.
- Setting team goals and building team spirit can help in all three areas of morale and provides the opportunity for individuals to ask for help in their own performance from others team members.
- If negative morale is widespread, organizations need to look to their hiring process and what can be done to improve the fit of the individual to the specific job competencies. This works to help ensure that employees can become competent in their jobs and feel good about their work. In addition, work design to improve job enrichment may be needed for individuals bored with their work.

D. Performance management and developing the individual

A very effective way to work on developing individual performance is through a well-designed performance management system. This is one of the most common organizational components, with close to 90% of organizations using this method to review performance. Performance management has great potential for developing individual as well as team performance because it can serve as a primary link between organizational/team goals and those of individual employees. In Chapter 14 of Part III we discuss this performance improvement method in detail. Here we will provide an overview of what is needed to make this initiative effective for developing individuals.

It should be noted that many performance management systems are judged by their users to be ineffective.[3] Experience shows that the reasons have to do with not following guidelines for construction and use of a performance management system.

1. Clear statement of organizational/team goals is a prerequisite for performance management. That way it is possible to set individual performance direction and have an objective basis for evaluating individual performance.
2. There should be clear identification of the core competencies needed for the job being managed. This is one of the most difficult guidelines to meet, but it is achievable if the organizational leadership is dedicated.
3. The performance management system should include a collaborative method for development of individual performance goals and Key Performance Indicators.
4. Another very difficult guideline for organizations to achieve is to have and use methods for regular feedback to employees on their performance. Accomplishing this guideline is difficult for a number of reasons include managers' anxiety about their ability to coach and counsel and the increasingly rapid pace of work in today's organizational world since time is at a premium.
5. The organization should provide training and development opportunities for enhancing the knowledge and skills of the employee which are supportive of his/her achieving a high level of performance.

A review of a brief case discussed in Chapter 4 will illustrate good use of many of the guidelines stated above.

Organization: A pet food sales team

- Products and Services
    dog food
    cat food
    specialized pet foods, litter

- Markets
    pet super stores
    regional chain stores
- Primary Issues
    the need to clarify their goals and actions in support of corporate strategy
    the need to build the team towards increasing cohesion and performance

The team had received corporate objectives and goals for the coming year. Having a good deal of independence in achieving their part of the corporate strategy, the team leader decided that the group should develop a strategic plan of its own, remembering that their own strategy should support that of the overall corporation. The team concluded strategic planning by developing a group of strategic objectives aimed at developing team performance. Following that, the group decided to drive the strategic objectives through each of the subteams within their larger team. They wanted each subteam to have accountability for parts of the strategic objectives and for individuals also to include specific goals in their individual action plans. They felt the need to conduct a process to cascade these objectives down to the individual level, rather than leaving it strictly up to the subteams and their leaders and employees. Busy schedules dictated that the team be involved in making sure the process of goal setting was completed by all individuals in their group.

After each member of the leadership team developed plans for his or her subteams, the total group met once more to coordinate all the efforts involved in their group plan. The steps they followed were as follows:

- Each team member listed his or her individual action plans which included both strategic objectives and the goals for his or her daily subteam operations.
- To the extent appropriate, the team members also discussed the goals and activities of the employees they managed where these goals and activities had significant impact on those of the group.
- The total team reviewed and discussed individual action plans looking to coordinate efforts, identify gaps in what needed to be done, and eliminate duplication. Further conversations with the Pro*Visions leader were scheduled for detailing of plans and identification of support needed by the managers and their subgroups.
- The leader also identified training and development services needed for his team members and those they managed. To a significant extent, those training opportunities were tied to the immediate performance objectives developed by this total team planning process. This included the installation of hiring tools and feedback on management style.

E. Core competencies and developing individual performance

Developing individual performance is helped significantly by recognizing that people are influenced in their job roles by what is occurring in their personal roles. While some employees are private and keep things to themselves, others appreciate and benefit from organizational support in areas providing personal benefits such as learning better communication, goal setting and time management skills that include but go beyond application at work. Others appreciate learning methods for improving personal finances, and make use of employee assistance plans.

But realistically, organizations provide these personal support programs primarily because they hope that a successful or "less troubled" personal life will lead to better job performance. But to help insure better job performance, organizations need to know what it takes to perform well on the job. What are the core competencies for successful performance for the various positions? Very briefly stated, knowing core competencies have the following benefits.

1. They provide the basis for rationalizing the hiring process so that people who have the core competencies for each position can be identified.
2. Identification of core competencies can drive the training and development plans so that it provides help to those seeking to improve performance through developing these competencies.
3. Core competencies provide a basis for setting performance goals at the beginning of the performance management cycle, and then evaluate performance at the end of it. They also provide target subjects for coaching and counseling individuals during the performance management cycle in order to develop peoples' abilities at the core competencies.

David D. Dubois defines job competency as the "underlying characteristic of an employee (i.e., motive, trait, skill, aspects of one's self-image, social role, or a body of knowledge) which results in effective and/or superior performance in a job."[4] The wisdom of this definition is that while it allows for the fact that the requirement for specific competencies come from the nature of the work, (job tasks and activities, roles and organizational environment and issues), the competencies lie within the individual.[4] The key point here is that once job competencies are identified, they can become the basis for hiring and then developing individuals toward those competencies.

F. Conclusion: Individual learning and performance improvement

As noted at the beginning of this chapter, all performance improvement points back to developing individuals, at least in part. Individuals ultimately must learn how to make the strategy happen or make the new structural design work. Or they must learn how to sell the new product, and manage the changed production process. Those of us working in performance development often get enthralled with new performance improvement interventions and forget a basic truth. When the interventions work, they work because human beings have the attitudes, the motivation, the knowledge and skills, and whatever other competencies are needed to make them work.

G. Suggested action steps for organizational/team leaders

1. Identify the common personal issues that have an impact on performance in your organization/team. (Financial problems, chemical abuse, etc.) Take actions to establish programs as needed.
2. Review any available data you have on morale in your organization. If data is lacking, conduct a brief morale survey. Decide what areas of morale need work and what sources of poor morale exist. Take action to eliminate or reduce those sources.
3. Review your performance management system. Is there sufficient goal setting, sufficient feedback on a regular basis regarding performance? If not, take action to build those performance management components.
4. Make sure that your understanding of the job competencies for core jobs is complete and up to date. If not, action is required.
5. Make sure that the core competencies are used in your hiring process.

End notes

1. Smither, Robert D., *The Psychology of Work and Human Performance*, Longman, New York, 1998, 217.
2. Sigelman, Carol K. and Shaffer, David R., *Life-Span Human Development*, Brooks/Cole Publishing Co., Pacific Grove, 1995.
3. Cummings, Thomas G. and Worley, Christopher, G., *Organizational Development and Change*, South-Western Publishing, 6th ed., Cincinnati, 1997.
4. Dubois, David, D., *Competency-Based Performance Improvement: A strategy for Organizational Change*, HRD Press, Amherst, 1993, 9.

chapter eleven

Training, development, and education: Helping people learn for performance

A. Introduction

Training, development, and education; while these are distinguishable performance development initiatives, they share a common purpose. Their basic purpose is to provide participants with concepts, skills, and perspectives which can enhance performance, now or in the future. In various ways, these three types of learning initiatives seek to inform us, to provide us with concepts with which to work, to teach us skills necessary for performance, and to encourage positive attitudes and motivation so we will use the concepts and skills successfully.

B. Distinguishing training, development, and education

The three terms are used in various ways in different organizations. But so much effort and resources are put into the three areas that clarifying what each means can help organizational leaders in deciding how and where to use these performance improvement initiatives. The organizational use of these interventions is enormous. In early 1998, ASTD estimated that all U.S. organizations spent a total of $55.3 billion on training in 1995.[1] This huge amount may not include the salaries of participants in training, or development, and educational programs such as support for college tuition or executive development seminars.

1. *Education*, in terms of current usage in organizations, is the vaguest of the three terms. Most commonly, it is used in two ways. First, education is used to mean anything an organization does, including job training, executive development, and tuition support for college courses. Secondly, the term is used to mean an organization's support for its employees attending college.

In this case, training and development is then seen as everything else the organization does.[2]

Many organizational/team leaders have found the following use of the term *education* to be helpful in planning their learning programs. Education includes:

- methods and programs for learning knowledge and concepts believed to be useful for the long term performance or career advancement of the participant.
- learning about performance programs/initiatives believed to be beneficial to the long-term performance of the organization/team. Examples are:

    a seminar for organization/team leadership on the characteristics and strategy for establishing a corporate university.

    a presentation to leadership on the approaches, costs, and benefits of process re-engineering.

2. Training and development can be more easily defined. Identity/Role theory will help here, discussed in Figure 10.1, reprinted below as Figure 11.1.

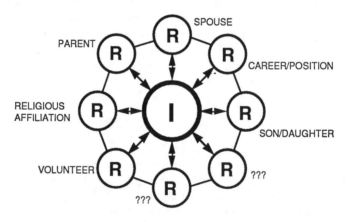

Figure 11.1   Identity/Role Theory.

Training is teaching people knowledge, skills, and attitudes about the roles they play, or will soon play, in the organization/team. Examples are as follows. A sales person is taught techniques for closing business, prospecting, or converting resistance to the purchase by the customer. Managers are taught the importance of and techniques for goal setting. Customer service providers are taught how to recognize and deal with various types of hostility in customers.

Development, on the other hand, is helping the person discover and/or better understand their own identity and how to get more out of who they are. As discussed in Chapter 10, it also includes belief in and efforts at developing one's personality, knowledge, and skills in multiple roles, motivation, and capacities. Review of a second simple model will further clarify the distinction and relationships between *Training* and *Development*.

Performance = Clear Goals + Good Attitudes
+ Knowledge/Skills + Motivation

Very simply, development has to do with work on the identity factors of a person's goals, attitudes, and motivation. Training focuses for the most part on improving knowledge and skills related to one's job roles.

C. The three learning initiatives and performance improvement

A number of excellent books have been written recently on the critical connection between corporate strategy and training, development and education for an organization.[3] The devil however, as they say, is in the details. Specifically, there are at least two major questions here.

1. Which of the knowledge and skills areas are most urgent for reaching our strategy? What do we need to learn to make the strategy happen?
2. What learning initiative best fits with other performance improvement initiatives we are using such as process re-engineering, organizational restructuring, etc.?

This section will provide an overview of important linkages between learning initiatives and other performance improvement initiatives, including strategic planning. Chapter 12 will add specifics by using a number of brief cases to illustrate those linkages.

D. Learning initiatives, strategic planning, and other improvement initiatives

1. *Strategic planning* and training and development are connected in two separate but related ways. First, the best benefit to the organization/team involved in strategic planning is not only to get a written strategic plan, but to learn how to do planning. The first time for strategizing by the top team, may require a facilitator from outside the group. But if training the team in how to do strategic planning is intertwined with the process of actually developing the strategy, the team becomes more independent and more knowledgeable of the process it has experienced. The team learns to plan.

2. *Improving the hiring/selection process* should often involve a number of performance improvement interventions including the following:

- defining job competencies for significant positions
- identifying profiling instruments, skills tests, or attitude surveys depending on the position
- conducting a normative study to identify which factors on the profiling/attitudes/skills instruments are predictive of job success

Training and development also plays a major role in a well-designed hiring/selection improvement project. Areas where training and development can be very beneficial include group learning on what the profiling/attitudes instruments measure and how to best use them in the hiring process, training on effective interviewing, and education on legal limits in the hiring/selection decisions.

3. *Restructuring and/or process improvement* is often done by organizations, especially those in the business world, with little attention to training and development. Marching orders are given to those responsible for the restructuring, or to those being impacted by it once the decisions are made. The same is often true for process improvement. For example, a consultant or internal process improvement expert, charts the steps and decision points in the identified process to be improved, then provides a report suggesting changes, many of which are installed with little discussion or explanation.

But training and development can also be very helpful in these performance improvement interventions. When restructuring occurs, education on the reasons for the changes, the challenges and strengths that might be predicted from new interpersonal work relations, and team building can be very beneficial at making the changes work.

With process improvement, a facilitation process is usually better than a consultative approach. Facilitating a process by which those involve learn how to re-engineer the process, and then do it themselves, usually produces a better quality and a more acceptable outcome. We will illuminate this approach more fully in the following chapter with a specific case.

4. *Team building* occurs in many forms, some of them involving training and development. Other types of team building interventions include diagnosis of the team functioning and discussion of team management skills. Training and development of teams helps improve team functioning by providing the opportunity for team members to learn model(s) of effective performance, learning and practicing useful skills in team work, and learning techniques for the team to self-management. Some training on team member roles is also beneficial to building teams.

5. *Performance management* provides multiple opportunities for training and development which are frequently ignored by organizations. The ineffectiveness of many performance management programs is at least partly due to insufficient training of program participants.

When a performance management process (often called an "appraisal" process) is first designed, much anguish, lost time, and ineffectiveness in using the process will be avoided if people are trained in a number of topics. Specifically, trained in these programs should include:

- What the process involves, its steps, and various roles of manager and employee
- How managers can best provide feedback, set goals with the employees, and help the employee identify their development needs and opportunities
- Methods the manager can use to provide feedback to the employee during the performance cycle
- Suggestions for employee perspective on the performance management process and how they might get the most out of it

6. *Developing individual performance*, discussed in the previous chapter, has many connections to training and development. In today's organizational world, with fewer middle managers and with increased demand for speed and effectiveness in dealing with customers and clients, individual employees need to know how to self-develop and self-manage. This is one of the key points in Peter Drucker's recent book.[5]

7. *Training and development* is a stand alone performance development intervention when individuals need additional knowledge and skills for job performance, whether or not another improvement intervention is under way.

E. Characteristics of effective training and development

If training and development is a valuable performance improvement intervention both as a stand alone activity and also to support other interventions, then what does it take to make sure these efforts work? Why do so many training and development programs seem to be ineffective, and how do we overcome that ineffectiveness?[6] This section discusses nine characteristics.

1. Those cited earlier in this chapter who stress the importance of tying education, training and development to the strategy of the organization/team are absolutely correct. Again, the primary issue is to identify exactly what learning outcomes are needed most urgently and which ones best benefit the most important elements of the strategy. You can't do everything at once to improve performance, but in today's competitive world, you must always be working on something. The best approach to identifying learning required strategically is to make it part of the strategic plan.

2. The role of dedicated and persistent leadership support in making any learning initiative more than an academic experience, stressing its importance

to current or future performance, is also critical to success. These first two characteristics are connected. Dedicated leadership should start their support of learning by making sure the strategy is clearly defined and stated, and the strategically-required knowledge, skills and attitudes are identified in the plan.

As the well-written book, *Training for Impact*, states:[7]

Learning Experience × Work Environment = Business Results

It is natural for people to percieve learning as more worthy if it is connected to organizational strategy.

3. A well-defined and facilitated learning experience (education/training/ development) is an obvious but sometimes elusive requirement for helping make learning experiences have an impact on performance. This characteristic starts with the learning facilitators' dedication and persistence to doing what it takes to provide the requisite learning opportunities. The requirement for making learning effective at improving performance is not the sole requirement of the facilitator. Nor is on the job application solely the responsibility of the organization/team leaders. It requires collaboration between the leader and the facilitator. The facilitator should be dedicated to helping the learning participants see how the knowledge and skills can be implemented at work. The learning program should also provide activities that require efforts at that implementation.

4. Awareness of the role of individual identity, and how it impacts performance of even the most technical roles, is a critical characteristic of effective learning. Participants get weary of learning which appears to benefit only their job role performance. As stated above, seeing the ways to implement learning on the job is important, but participants are more motivated by learning that involves benefits for them in areas outside of their current work. For example, teaching sales persons how to confront resistance and sell the organization's products/services more effectively is received more readily when some of the examples pertain to learning to confront issues with friends or family. One participant in a project of the State Department of Family Services described in characteristic eight below was particularly impressed with discussion of "asking and listening" when she tried them on her teenage son and found they worked. She decided the techniques were worth using at work as well.

An additional brief case will help illustrate this point regarding how attention to personality, motivation, and attitudes can have an impact on the success of training and development.

Organization: Sales team: pet products

- Products and Services
  pet products

- Markets
  veterinarians across the U.S.
- Structure
  national sales director
  a small number of divisional sales managers (DSM) with numerous
    sales people under their direction
  approximately 50 fields sales people

Issue: The national sales director and at least some of the DSMs experienced a lot of interpersonal conflict. Causes of the conflict included disagreement over leadership effectiveness, between the national sales director, the divisional sales managers, and the field sales people. Survey data included that field sales often felt in the dark about policies and procedures. Communication and trust was at a fairly low level throughout the team.

It was clear from the beginning, however, that personal dislikes also played a part in the conflict.

The performance improvement initiative that was determined to be most promising was a leadership development program using the Achieving Manager Model of leadership discussed earlier.

This leadership model includes a requirement for open communication. This means that communication would have to include personal attitudes and biases, in addition to learning leadership/management skills. Once the need to focus on personal motivations and attitudes became known, a series of discussions occurred between various persons involved in the more personal conflicts. The open communication in the leadership model helped. The desire grew among participants to find some accommodation between the combatants. This led to improvement in interpersonal relations between three of the upper level team members. Effort was also directed at increasing the policy communication with field sales, and they were included in team problem solving efforts. Data indicated an improvement in the attitudes of field sales regarding communication and their involvement in solving specific issues affecting their job.

Identity issues such as attitudes towards co-workers, personal attitudes affecting those attitudes, and abilities at confronting and resolving interpersonal conflict, are never easy to improve. Interpersonal relations are, however, always present at work, with positive and/or negative results. Not confronting these issues where they exist means making the learning ineffective.

5. The use of spaced repetition is a key to learning sufficiently to make use of knowledge and skills. Figure 11.2 illustrates this point.

The vertical dimension represented by the above figure demonstrates retention of learned content. The horizontal dimension, from lower left hand corner to the right side, represents time. The point of the figure is that one hearing/seeing of content is quickly forgotten over time. The repetition of hearing or seeing that content increases retention over time. Unless

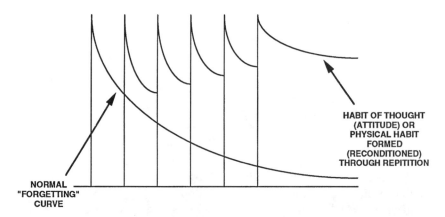

Figure 11.2 Spaced Repetition Model.

something is retained in the minds and memory of learners, it cannot have an effect on their performance.

Some practical considerations in daily training activities are important here. Many participants who regard themselves as quick learners are especially resistant to any activity involving repeating of old material. This is sometimes a major issue between facilitators and participants. Repetition and resulting retention determines how much the participants will be able to use the learning. Happily, in today's high technology world, it is possible to have learners see the same content through various mediums, increasing their interest and decreasing their boredom. Reading and watching videos on the same content is repetition with variety.

6. Effectiveness in training and development is enhanced by the use of a simple method which is also useful in on the job training. Very simply:

SHOW → THEY PRACTICE → DISCUSS → THEY REPEAT PRACTICE

Here the distinction between education and training and development discussed earlier in this chapter, is important. Education usually involves maximum attention to concepts and minimal attention to skills development. Training and development are much more skills focused; training pays attention to job skills and development focuses on personal skills. The key point is that knowledge of concepts or skills can only become part of our behavioral style through practicing those concepts and skills in the real world. While role playing in a training session can help in development of know how, practicing in the actual work situation is best.

7. Effective education, training, and development maximizes self discovery. What that means is that participants need to have multiple opportunities to:

- decide how the concepts being learned apply to them
- decide how to make use the knowledge and skills they are learning; plenty of time for setting specific goals and action planning needs to be provided

8. Education, training and development, should make use of as many of Donald Kirkpatrick's four levels of evaluation as possible.[8] Most learning facilitators are satisfied with "reactions" to the learning on the part of participants. As Kirkpatrick has shown, that is necessary but not sufficient for knowing if the learning will lead to performance improvement. The following is a fascinating case of training large numbers of people in a program offering many options for evaluation.

Organization: Division of Family Services/Boot Heel of Missouri

- Products and Services
  welfare cash benefits, child care, employment training and referrals, health care benefits
- Markets
  welfare recipients who were under legal pressure from Welfare Reform
- Structure
  area III of the Missouri Division of Family Services (DFS) headed by an area director and other central area staff twenty-four southern Missouri counties comprise Area III. each county delivers products and services to those eligible for welfare in their county areas
- Issues/Problems
  Welfare Reform of 1996 set legal time limits on case benefits to welfare recipients

The Welfare Reform actions by the federal government in 1996 have been thoroughly discussed in many publications and have been shown to have huge implications for welfare recipients as well as national policy and the economy of the United States.[9] Each state is obligated to meet the time limit responsibilities of the Welfare Reform legislation, though leeway in how they did that exists. Missouri took the initiative to support a massive training and development program for nine of the county welfare agencies responsible for making "Welfare to Work" happen. Over 250 participants were included in 48 hours of training and development spread over five months of effort.

The evaluation of the program included the following thrusts. First, reaction feedback from the participants, Kirpatrick's first level of evaluation, was obtained at the end of each module. Secondly, the content knowledge of a random sample of the 10 groups of participants, approximately 25 participants in each group, was measured in both a pre- and post-seminar setting. Some groups were given both post- and pre-seminar content tests. Some groups were measured only before but not after the seminar, and vice versa. As a control measure, the content learning of a few of the groups was not measured at all.

Evaluating changed behavior and improved results, Kirkpatrick's third and fourth level of evaluation, is by necessity a longer term project than the first two levels of evaluation. The basic legal requirement for the county Division of Family Services (DFS) offices is to increase the percent of welfare recipients who are involved in some type of job activity, training or work. The central DFS staff had the results of job involvement efforts for each county prior to the training. It is comparatively easy to evaluate the effects of the training through changes in these performance numbers after the five months of learning are completed. One word of warning is important. It is never possible to totally isolate the effects of a learning initiative on an organization from other factors that might have influenced performance. For example, in this training initiative changes in legislation or state regulations allowing the county offices more or less freedom in doing their work could affect performance independently from the learning experience. Still, improved performance is highly desirable in any organization, whether we can know exactly what caused it or not.

In the Boot Heel project, evaluating improved behavior was somewhat more complicated than the other areas of reaction, content learning and results. However, an organizational climate survey had been completed prior to the learning workshops for county workers and was then available for completion after the seminars were completed. This means there is the possibility of a pre- and post-evaluation using employee perspective on the behavior of employees in the county office.

9. The use of multi-sensory learning approaches adds to the effectiveness of training and development. It is well established that people learn differently; some learn better from seeing, some from hearing, and some from activity like writing or role playing. It is usually not possible to assess each participant's learning style, certainly not in a large program like the Division of Family Services initiative discussed above. That generally means that using multi-sensory inputs is important in all programs, allowing the participants to learn the way they learn best. The technical approaches available today, from power point presentations, to interactive computer based training, make the use of multi-sensory learning approaches easy to achieve.

10. The inclusion of methods for reinforcing and using learning back on the job is the ultimate test of whether the learning has worked and been worth the time and expense. In Part III of this book we will discuss the methods available for making learning permanent so that it can impact performance.

Here, it is important to take notice that reinforcement of learning must occur in the daily work of the organization if this initiative is to meet its ultimate test. Did the learning help those involved perform better than before?

F. Suggested action steps for organizational/team leaders

1. Make sure that your strategy has been well designed and communicated to those required to provide or experience learning. Help make the connection between strategic accomplishment and learning apparent.
2. Determine how important you believe learning to be for achieving the organization/team strategy.
3. Identify and begin actions on those areas where your support as a leader will enhance dedication on the part of learning participants. Keep in mind that this is important not just during the training/development, but after the structured learning experiences have occurred as well. Consistent reinforcement is the key.
4. Make sure that any learning facilitator involved with your organization/team has as a primary motivation helping people learn. Personal benefits will exist, but they must be seen as legitimate only as a result of effective facilitation and learning.
5. Whether you use learning facilitators from outside or inside your organization/team, make sure they appreciate and use methods and approaches that make the learning effective.
6. Make sure that your learning initiatives have methods for evaluation on all four of Kirkpatrick's measurement levels.

End notes

1. Bassi, Laurie J. and Van Buren, Mark E., *State of the Industry Report*, Training and Development, Alexandria, January 1998.
2. Blanchard, P. Nick and Thacker, James W., *Effective Training*, Prentice Hall, Englewood Cliffs, 1999.
3. For example: Blanchard, P. Nick and Thacker, James W., *Effective Training*, Prentice Hall, Englewood Cliffs, 1999; Mesiter, Jeanne C., *Corporate Quality Universities*, Irwin Professional Publishing, Burr Ridge, 1994; Brinkerhoff, Robert O. and Gill, Stephen J., *The Learning Alliance*, Josey-Bass Publishers, San Francisco, 1994; Robertson, Dana Gaines and Robinson, James C., *Performance Consulting*, Berrett-Koehler Publishers, San Francisco, 1996.
4. Cummings, Thomas G. and Worley, Christopher, Worley G., *Organization Development and Change*, South-Western Publishing, Cincinnati, 1997, 225.
5. Drucker, Peter F., *Management Challenges for the 21st Century*, Harper Business, New York, 1999, 142.

6. Many excellent books on training and development, including those cited in note 3, are concerned with improving training and development. Perhaps no publication has been as dramatic in making the point about the limited effectiveness of learning experiences as the nationwide headlines a few years ago from USA Today. "Big Lesson: Billions Wasted on Job Skills Training," *USA Today*, Wednesday, October 7, p. 1, 1998.

7. Robinson, Dana Gaines and Robinson, James G, *Training for Impact*, Josey-Bass Publishers, San Francisco, 1989, 109.

8. Kirkpatrick, Donald L., *Evaluating Training Programs*, Berrett-Koehler Publishers, San Francisco, 1998.

9. See for example: Schorr, Lisbeth, *Common Purpose*, Anchor Books, New York, 1997.

chapter twelve

Learning in support of other performance improvement initiatives

A. Transition

"There is nothing as useful as a good theory." This phrase, attributed to the renowned teamwork expert Kurt Lewin, is an illustration for the purpose of this chapter. What is contained here is a series of brief cases of organizations attempting to use learning initiatives to increase performance of their employees. Some of these efforts worked pretty well, others not so well. But the test of a good theory, such as the Characteristics of Effective Training and Development discussed in the previous chapter, comes from application and review of the results.

As noted earlier, training, development, and education are sometimes useful for performance enhancement in their own right. Sometimes the learning activity is seen as having indirect impact on performance. For example, if employees are provided programs to improvement of their personal finances, they may be better able to concentrate on their jobs. Other times the connection between what is to be learned and performance is very direct, as in leadership training, job skills training or customer service.

What tends to be ignored is the value of learning in performance improvement efforts that are primarily focused on other initiatives such as restructuring, strategic planning and organizational development. While an overview of the connection between training and development and other performance initiatives was introduced in Chapter 11, here we identify specific connections. The following cases are descriptions of training, development and education as critical supports for helping other performance improvement efforts work.

B. A comprehensive effort at performance improvement

Organization: A hinge company

- Products/Services
  manufactures hinges of all sizes and uses

- Markets
    distributors
    builders
    manufacturers
- Structure
    corporate headquarters in the Midwest
    plants located in three southern states
    wide network of distributors for all types of their products
- Problems/Issues
    the top managers saw a need for improved leadership skills and
      corporate direction

The leadership of this successful company realized that they needed to improve performance in a number of areas, though which areas was a matter of discussion and some disagreement. They recognized that part of the issues they faced were morale questions resulting from concerns about nepotism. The company is owned by two families, who also dominate the top leadership positions in the organization.

The decision was made to start performance improvement efforts with a leadership development program featuring feedback from the direct reports of the top managers. The focus was on improving management and leadership abilities and dealing where necessary with anxiety and distrust due to family domination of the firm. Over a number of months the leadership team met and reviewed their feedback, completing action plans for improving their manager-subordinate relationships. The leadership development program was so well received that it was used in a number of the plants as a follow up to the initial effort. The learning experiences achieved in the seminar were tied to specific organizational development actions in the corporate office as well as in the plants.

Strategic planning was the second performance improvement initiative chosen by the hinge company. The top team learned how to do strategic planning by constructing its own plan and at the same time discussing the process it was using. The result was a thorough strategic plan including many objectives such as improving their marketing plan and increasing the quality and efficiency of the production processes in their plants. The strategic plan became a central focus for many spin off projects throughout the next few years.

The third performance improvement initiative undertaken by the company was the creation and installation of a performance development system (PDS). The top team of managers was involved in making decisions regarding the basic characteristics of the PDS, and the final design was done by outside consultants. All of the managers and employees who were to be involved in roll-out of the performance development system experienced a learning seminar where they came to understand how the PDS worked, what their roles in the system were, and how they might best perform those roles successfully.

Two measurement factors were especially important during development and installation of the PDS at the hinge company. First, the goals they set during the planning period between manager and employee was to include attention to things that needed to be done to support the strategic objectives discussed above. This was a classic example of cascading goals.

Secondly, the company began to make more systematic use of Key Performance Indicators (KPI). KPIs are somewhat easier to use in a manufacturing setting compared to a service industry. The company connected their use of KPIs to efforts at process re-engineering and manufacturing quality improvement.

The company has done a number of things in recent years to work at performance improvement. They have developed a strategic plan and communicated it throughout the company. They have worked at achieving the strategic objectives they developed and have done well in most areas of effort. They have developed and installed a performance development system with their employees, making an effort to have the specific goals build into the individual performance plans be connected to the strategic objectives. The leadership of the organization has also taken measures to improve their production processes. What is most significant here is that they recognized early that they needed good leadership to make these performance improvements happen, and so leadership development was their very first effort.

Education, training and development was a part of every phase of this comprehensive performance improvement effort, including; developing a strategic plan, leadership development, and establishing a performance development system.

C. Learning and strategic planning

Strategic plans can be developed any number of ways, including use of external industry specialists who conduct interviews and then write and present the plans. Scottsdale Securities is outlined below (it was discussed briefly in Chapter 4); it is a success story in using learning experiences to develop and manage a corporate strategy.

Company: Scottsdale Securities, Inc.

- Products and Services
  on-line discount broker
  broker assisted trades as needed
- Structure
  approximately 100 corporate personnel
  approximately 106 branches spread throughout the U.S.
- Primary Issues/Problems
  lacked shared strategic vision

> CEO who was the founder and primary owner had at least a partial vision of what he wanted to accomplish, but it was not fully developed and not completely understood and/or accepted by his corporate and branch managers

- Actions
  > conducted a thorough strategic planning process with the twenty or so top managers at the corporate level

The leadership team not only developed a strategic plan with 19 objectives, but they made major progress at learning how to do strategic planning in the future. At the end of the first 14 months after completing the plan, they had achieved a number of the objectives and made progress on others. The objectives included creating new products and services, major shifts in their marketing strategy and technical back up systems for their on-line equity trading systems.

It was interesting to see people at Scottsdale, especially those at the top of the organization, begin to use concepts of strategy versus operations discussed in strategic planning and in the model for organizational success.[1] That learning of concepts helped them in discussions about may organizational factors including operational improvements that they came to discuss and evaluate in terms of strategic benefits or consequences.

The training and development the 20 Scottsdale managers experienced while developing their strategic plan was not the end of their drive to learn about performance improvement. One of the strategic objectives, as mentioned in an earlier chapter, was the requirement for establishing a training function. They did this. Another strategic initiative, also mentioned earlier, was their desire to experience training and development in leadership/management. They accomplished that objective as well, and a number of the systems and procedures for managing more effectively became a part of how the managers performed their management roles.

The Scottsdale strategy also included a specific objective on opening a number of new branches throughout the United States. The leadership came to realize that this required improvement in their hiring process. They decided to work on improving hiring by adding personality profiling to the information they used in making hiring decisions. They also decided to have a normative study done, the use of high performers in the branches as a way of deciding what types of personality profile scores to look for in the future. Finally, as of the writing to this chapter, the firm is undergoing an update of its strategic plan. They have learned that strategic direction for the organization are a key to performance and performance improvement.

D. Learning and teamwork

One of the most common efforts at performance improvement is team building and education on teamwork. This comes from a number of factors including reduction in work forces in many organizations and the belief that productivity and synergy can result from good teamwork. When teams are used at the right time and for the correct reasons, these benefits, as well as others we discussed earlier in this book, do occur. But leadership judgment on when and how to use teams is required to maximize benefits and minimize costs of teamwork.

Team building can be a part of other performance improvement initiatives, or it can be a stand alone effort. For example, in strategic planning when the leadership team works together to define and achieving the strategic goals of the organization, teamwork is developed. The same is often true for working with an intact team such as a sales or customer service group. When people work together to identify, solve problems and meet goals, they are performing as a team.

In Chapter 9 we discussed the organization listed below in the context of improving leadership performance. The leadership team, however, also decided to work on their team functioning and the relationships between team members.

Organization: Ranken Technical College

- Products/Services
  technical education
- Markets
  adults
- Structure
  central leadership team of about eight people
  president and vice president with strong personalities
  two faculty groups: general studies and technical studies
  staff
  about 2000 students
- Primary Issues/Problems
  evidence from an organizational culture survey and discussions with some influential leaders in the organization suggested that power was overly-centralized with too few decisions made by people who had to implement them
  some major inter-personal conflicts existed between the two top leaders and others in the organization. There was belief that performance was being hampered, in part because of the time and focus spent on unproductive disputes.

As discussed earlier, the leadership group of the college decided to go through a development process in which they learned the High Achieving Manager leadership model. During that project, they also received feedback from at least three direct reports using survey instruments that duplicated the original project. Among the specific changes that occurred in leadership

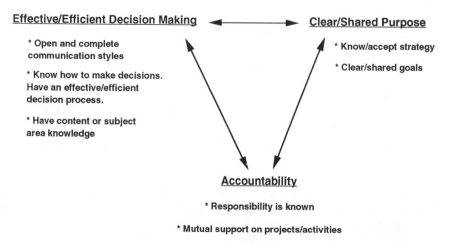

Figure 12.1   The Model of Effective Teamwork.

style was the decentralization of decision-making power and the empowerment of faculty and other teams to make those decisions consistent with their expertise and interest.

This leadership group had originally started their effort at performance improvement with a team building project in which they sought to identify why their interpersonal conflicts were so severe. The model in Figure 12.1 helped the team begin to identify issues where their team clearly needed development.

During a series of meetings, the team engaged in a good deal of team self-assessment. The conclusions they reached, using the Model of Effective Teamwork presented in Figure 12.1, were as follows.

1. The team tended to agree on overall organizational strategy regarding the mission and direction of the college.
2. There was some team disagreement on specific goals, such as how much emphasis to put on night and weekend programs, but generally there was agreement on specific goals.
3. The tensions between team members were principally in the other two areas of effective team work — accountability and decision making and communication between members. At least some of the team members feared retribution from being held accountable for specific and measurable goals.

Most significantly, many team members believed that communication between the members was fraught with private agendas, suspicion, and self-protection. In a mature judgment, the group initiated discussion of the most fundamental of issues about top teams. The issue discussed was how their own interpersonal issues were affecting the rest of the organization. It was at that point that they realized the need to work not only on their own team functioning, but to also undertake leadership development to include people with whom they worked directly. That is when they began to make the decision to undertake the leadership development program. The team was beginning to work better together to improve performance.

E. Knowledge and restructuring the organization

In today's organizational climate, with strong motivations to be competitive, reduce costs, and build equity value in for-profit organizations, restructuring of organizations is everywhere. This is frequently done for reasons other than, or even in conflict with, performance of the resulting organization/ team. Still, whatever the motivations for reorganizing, using the knowledge of those most familiar with the functions and people involved in the units being reorganized is wise. The following case demonstrates a useful process for planning reorganization.

Organization: Pet food supplied to veterinarians and breeders

- Markets
  veterinarians
  pet food stores
  leading breeders
- Products/Services
  pet food and related products
  recommendations from breeders and influential veterinarians
- Structure
  three teams focused on different markets inter-linked in activity at times, but often functioned independently; all were field selling groups. Breeders worked with pet shows and interacted with leading breeders to obtain recommendations for company products.
- Problem/Issue
  for some time there had been a recognition that better coordination between the teams might add to the selling and marketing success of the three groups. There was an issue of whether the structural separation between these groups sometimes hampered the marketing/selling process.
  executives got a report which showed that combining the sales forces could result in significant increase in the number of sales calls being made by the same number of sales people.

The leadership of the three teams was asked to develop recommendations for how to restructure the three groups so as to combine the two sales forces and relate the breeders group to that new combination. Each of the teams had developed a written strategic plan identifying their markets, products to be sold to the markets, selling/marketing strategy. During the process of creating these strategies, they had learned a good deal about how to develop a strategic plan. The decision on the part of the leadership of these groups was to use strategic planning and team decision making as the foundation for making the restructuring recommendations to higher level executives. They had learned that the structure they recommended should follow the business processes which could be identified in the strategic plan. They believed that the strategic approach could better maximize marketing/selling performance then re-organizing in a strategic vacuum.

The team completed a strategic plan for the three formerly independent groups now needing to merge into one. They then developed a structural recommendation to meet that strategy. The total plan was presented to executives at the firm accompanied by rationale for the group's decisions.

F. Learning and developing leadership

As discussed earlier in sections on leadership performance, finding and using a model of effective leadership/management is an efficient way to develop leadership style.[2] But many leaders/managers have had years of experience which leads to the development of habits and attitudes about leading/managing which resist change and improvement.

Sometimes managers need to get feedback from those with whom they work identifying their deficiencies as leaders. Without that, they often continue in the same old way, doing the same things, and getting the same poor results. The case below, restated in a different way from an earlier section, clarifies the issues and possibilities for performance improvement in this situation.

- Markets
    people on welfare who were under legal pressure from Welfare Reform
- Structure
    Area III of the Missouri Division of Family Services
    headed by an area director and other central area staff
    over twenty southern Missouri counties formed into Area III
    each county delivered products and services to those eligible in their county areas
- Issues/Problems
    Welfare Reform of 1996 set legal limits on allowable terms for welfare dependency

The organizational climate results completed by a high percentage of workers and managers in each county was presented to the county groups in the learning sessions. It was clear that in at least a few of the counties there was a great deal of unhappiness about how leadership and management was occurring. Once that lesson became clear, the issue remaining was to learn and make use of more effective leadership and management approaches.

G. Suggested action steps for organizational/team leaders

1. Make a list of any performance improvement initiatives planned or currently underway. Are sufficient learning opportunities connected to these initiatives to make sure they are as effective as possible?
2. Is there any reorganization planned in your organization/team? What kind of learning is needed to make this very common initiative effective?

End notes

1. This was discussed more fully in Chapter 3.
2. See Chapter 9 for discussion of developing leadership/management teams.

part three

Evaluating and stabilizing performance improvement

chapter thirteen

Factors affecting performance improvement stability

A. Introduction and linkage

To this point we have covered the following major topics.

- Part I discussed the powerful need in our society for organizational performance improvement and listed some of the less desirable motivations that lead organizational/team decision makers into initiating efforts at performance improvement. In addition, Part I provided:
  A definition of performance
  A discussion of the importance of leadership providing a thorough strategic vision for the organization/team
  A discussion of approaches for the identification of performance gaps
  A model of planned change for identification and management of performance improvement where and when it is needed
  A discussion of the role of performance measurement
  A discussion of the benefits and characteristics of goal setting and its role in performance
  Numerous cases to emphasize how these elements can be achieved

Part II began with a discussion of the importance of building a culture and structure for a learning organization. In addition, Part II provided discussion of a series of performance improvement interventions, including:

- Selection/Hiring
- Strategic planning and cascading goals
- A model for effective teamwork and team building
- Developing leadership teams
- Technologies and approaches for performance improvement at the individual level

- Training, development, and education as stand alone performance improvement interventions
- The use of training, development, and education to support other performance improvement interventions, a deficiency in many organizational efforts at improvement

Part III pays attention to one of the most difficult aspects of performance improvement, evaluating and stabilizing improvement after the intervention for that improvement has been completed, or is in full swing. For clarity of communication, the factors influencing the ability to evaluate and stabilize performance improvements is divided into the following areas:

- The Critical Role of Leadership
- Characteristics of the Organization and its Culture
- The Personalities, Values and Attitudes of the People Involved
- The Characteristics of the Performance Improvement Intervention

B. First factor affecting evaluation/stabilization: The critical role of leadership

Direct experience with hundreds of organizations, combined with writing from many authors in performance improvement, mutually supports the virtual certainty of the following principle:

> The dedicated and persistent support of organization/ team leadership is often the most important factor in making performance improvement happen and stabilizing progress once it is achieved.

While diagnosis of performance gaps and their causes is also very important in improvement stability, that diagnosis will usually lack important data and sufficient insight without the dedication and involvement of leadership. In addition, how seriously the performance intervention is taken and whether effort is maintained after the initial enthusiasm is also dependent on leadership's appropriate involvement. The responsibility leadership of organizations/teams should take to make performance improvement work is best understood through use of the Model of Planned Change, originally discussed in Chapter 3 (Figure 3.1) and presented again in this chapter as Figure 13.1.

Entry and contracting is the first stage of this model for decisions regarding management of performance improvement. This primarily involves formation of an agreement with organizational/team leaders on a process for answering the following questions:

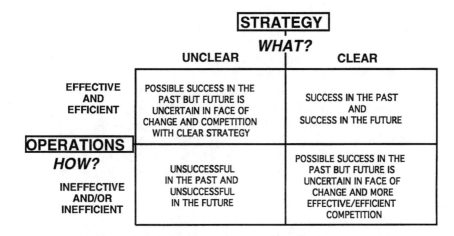

Figure 13.1 The Model of Planned Change.

- What does excellence in performance mean to us specifically?
  Quality and quantity of production of our products/services
  Effective marketing/selling and/or efficient marketing/selling
  Mutually beneficial and satisfactory customer service
  Effective and efficient distribution of products/services
- In the areas where excellence of performance is important to us, where do we have specific goals and performance measures (KPIs, Standards), and where do we not?
- Where we know or suspect we have performance gaps, how should we proceed to diagnose the degree and causes of those deficiencies?

Designing or supporting a process to answer these kinds of questions can strike anxiety in the hearts of even the most courageous organization/team leader. The most common fear of leadership is included in the following reflection heard from many persons:

> "What if I am a major cause for the performance deficiencies we have?"

Willingness on the part of leaders to support a diagnostic process where the causes of performance deficiencies are honestly identified requires a great deal of dedication and confidence from those in leadership positions. If internal or external performance improvement consultants are also involved, trust is often a greater issue. Full candor about what is working and what is not, and willingness of all participants to discuss the feedback is the key. Leaders set the tone for this phase.

The second stage of the Model of Planned Change, Diagnosis, involves technical procedures of data collection and interpretation of what the data says about performance. The most critical role of leadership is during feedback and discussion about what the data seems to say regarding performance, gaps, and causes of those gaps.

There are two very important requirements for leadership in this stage. First, they must be open to the results of the diagnosis. This does not mean blind acceptance of the interpretations. Interpretation of data in areas of human performance are always based on judgment on the part of the interpreter, and is therefore not to be taken as gospel, no matter how impressive the credentials of the person making the judgments. But willingness of organization/teams leaders to understand and consider the interpretation regarding causes of performance results and causes of performance gaps is a must if serious improvement has any chance of occurring.

Here there is value for the role of performance improvement experts, especially those with no on-going responsibility for the organization. Where possible, the performance experts, whether from inside or outside of the organization, should not have been a part of the leadership team with emotional ties to the results of the diagnosis. A performance consultant from the human resources function of a large organization, for example, is one possible source of judgments about performance that may not be influenced by emotional attachments and personal agendas. Outside experts are another source.

The second requirement for leadership during diagnosis is to be fully involved in deciding about whether a performance improvement intervention is needed, and if so, what it should be. A related decision is the involvement of leadership in any selected performance improvement intervention. Clearly, the leadership needs to be deeply involved if the selected performance improvement intervention is strategic planning. But the role of leadership in other interventions that might be selected is a matter of answering the following question.

> "What involvement should the leaders have in the performance improvement intervention in order to maximize the effort directed towards that intervention and the improvement results it achieves?"

The third stage of the Model of Planned Change, the Intervention itself, requires varying degrees of leadership involvement. The following three scenarios clarify the options leaders have in deciding on their best role during the intervention process.

1. The organization/team leader(s) clarify their beliefs about what performance the team needs to accomplish, what current performance is good and where it appears some improvement is needed. The leaders' support for the work the team is planning to do for improvement is stated clearly, along

with any resources the leader can offer to make the intervention happen. If the leader is not heavily involved in the functions of the team, more participation in the development of the team may not be needed or wise.

However, if the leaders are themselves directly involved in daily functioning of the team, then staying out of direct involvement during the intervention is unwise. First, it sends a "de-motivating" message to the group when the leader does not stay involved. For example, if this is a sales team, the leader and the members of the group may need to do work of designing new market-to-product connections, identifying factors to measure in selling reports for selling activity, or experiencing sales training to learn better closing techniques.

2. In this second scenario regarding leadership involvement in the performance improvement efforts, the leadership of the organization starts with the support identified in scenario one above. However, the process for performance improvement is designed to generate further information and recommendations on needed improvements. These recommendations for further improvements require that leadership stay involved on a timed basis so as to make decisions about which recommendations to use and how to do it. The following brief case, initially described in Chapter 11, illustrates this point.

Organization: Division of Family Services/Boot Heel of Missouri

- Products and Services
    welfare cash benefits, child care, employment training and referrals, health care benefits
- Markets
    welfare recipients who were under legal pressure from Welfare Reform
- Structure
    Area III of the Missouri Division of Family Services
    headed by an area director and other central area staff
    over twenty southern Missouri counties formed into Area III
    each county delivered products and services to those eligible in their county areas
- Issues/Problems

You may remember that Missouri took the initiative, prodded in part by the dean from a college of health and human services, to undertake a massive training and development program for nine of the county welfare agencies responsible for making "Welfare to Work" happen. Over 250 participants were included in 40 hours of training and development spread over five months of effort. As part of the pre-work, all of the participants completed a brief organizational climate assessment of their county. This included assessment of organizational clarity, communication, decision making and allocation of responsibility in the county organizations.

During the training sessions, participants were organized into small teams by county. Each county team took the data regarding their county that came from the organizational climate survey and used a problem solving process to decide how to improve the climate in their organization. Recommendations for improvement ranged from rearranging work load, to new methods for communicating the frequent policy changes, to developing the leadership and management style of directors and supervisors. Many of these recommendations were supported by the climate survey data and presented well. The area director, responsible for the entire group of participating counties, stayed involved throughout the many months of problem solving in the training program. Her approach was to evaluate each recommendation, and where possible, implement a large percentage of the recommendation by working with the director responsible for the county affected. Predictions are that this long term dedicated involvement from the top executive in Area III of the Missouri Department of Family Services will pay performance dividends.

3. The third scenario that clarifies the options leaders have regarding involvement in performance management initiatives is exemplified by the case of a technical college discussed originally in Chapter 9. As a brief review:

Organization: Ranken Technical College

- Products/Services
    technical Education
- Markets
    adults
- Structure
    central leadership team of about eight people
    president and vice president with strong personalities
    two faculty groups: general studies and technical studies
    staff
    about 3000 students
- Primary Issues/Problems
    evidence from an organizational culture survey and discussions with some influential leaders in the organization that power was overly-centralized with too few decisions made by people who had to implement them
    some major interpersonal conflict between the two top leaders and others in the organization such that performance was probably being damaged in part because of the time and focus spend on unproductive disputes

The leadership group of the college decided to go through a development process in which they learned the High Achieving Manager leadership model and were to take actions to move closer to that model. They received feedback from at least 3 direct reports using survey instruments that positioned them in relationship to the High, Average, and Low achieving managers. They then set goals and action steps to improve their style to become more like the High Achieving model. This included a series of meetings with the direct report to discuss specific steps in leadership/management improvement. The project lasted about six months.

The CEO of the college, having a dominant personality, consciously worked to remain less assertive about results from and actions to fix the organizational culture survey. While it was clear that he wanted improvements in culture, he was not going to dictate changes. He wanted other leaders to take responsibility as well. He was fully involved in group decisions as the team discussed improved leadership styles and moving the power of decision making further down into faculty and staff groups. He also received feedback from his own direct reports and fully participated in working with those direct reports, as well as others, to improve his leadership style. This leader was totally involved in the intervention.

The fourth stage of the General Model of Planned Change, Evaluation and Stabilizing Performance Improvements, requires the power of leadership position in making sure that a frequent deficiency in improvement efforts does not happen. That deficiency is the dissipation of the progress after the effort begins to wind down.

There are a number of ways that leadership can be involved in stabilizing performance progress. One is through the power of recognition. There is plenty of evidence that many managers in the United States operate on the "exception principle of leadership," that is, they only pay attention to problems and tend to ignore positive occurrences. This is not only poor leadership style, it is ignorant of a basic principle of human behavior. That basic principle is that rewarded behavior tends to be repeated.

To his credit, the CEO of Ranken Technical College, discussed above, worked hard to overcome his personality-driven tendency to ignore the positive and focus on the negative. Sincere recognition and reward for performance progress is not only good ethics, but good motivational psychology.

Leaders should also make sure that areas for performance improvement are measured and evaluated in performance management systems. We will discuss this more fully in the next chapter, but an example here will help make the point clearer. If a bank has decided that referring clients for potential sales between departments, (loan officers referring to credit card services) is desirable, then education and training about those procedures and goals needs to be completed. After that original attention to the performance improvement project has occurred, however, it would be wise for the leaders to make sure that their performance management system includes goals for and evaluation of the referral activity. This combines recognition of the new performance with rewards for making it happen.

C. Second factor affecting evaluation/stabilization: Characteristics of the organization and its culture

1. Organizational culture is one major factor in how well an organization/team does at stabilizing performance improvements once they have occurred. Organizational culture can be defined as shared beliefs, values and behavioral norms and the resulting visible aspects of the organization such as structure, systems, methods and procedures, work rules, and physical trappings of equipment and technology machines.

It has been noted that the interest in organizational culture, a term borrowed from cultural anthropology, is thought to have two major areas of impact which I have underlined in the following quotation. "There is considerable speculation and increasing research suggesting that organizational culture can improve an organization's <u>ability to implement new business strategies</u> as well as <u>to achieve high levels of performance.</u>"[1]

Which dimensions of organizational culture most influence the abilities to adopt new strategic directions and/or achieve high levels of performance are matters of controversy. Experience, however, suggests strongly that the following dimensions are most significant.

- Goal setting, measurement, and results focus
- Open communication and a willingness to discuss even the most difficult topics
- Valuing leadership that is oriented to both results and people development and strong support for at least some of the current organization/team leaders

Without a strong orientation towards goals, measurement, and results, no one really knows the status of performance and where the gaps are. Without the willingness to openly discuss even highly sensitive performance issues, difficult topics and issues are not confronted, and things go on as usual. Without strong appreciation for a leadership style that both values results and people, the High Achieving Manager discussed in Chapter 9, either people are de-emphasized and their motivation declines, or everything is comfortable and nothing improves. A brief case will clarify this point.

Organization: A medium sized printing company

- Products and Services
    printing with specialization in corporate annual reports
- Structure
    new CEO after many years of leadership from a now semi-retired owner
    vice presidents for sales and finance
    divided into sales group and production group

- Issues/Problems
    There were major conflicts between production and sales concerning scheduling work and delivery times, a class production-sales dispute seen in many organizations. There had been attempts to solve the work scheduling problem before, but after some apparent agreement about actions to improve performance in this critical area, little happened. At least some of the sales and product employees did not believe that the leadership group would make them do what was needed in order to reduce scheduling conflicts and production errors and improve on time delivery to customers.

The sales and production leaders, prompted by their leadership team, experienced a team problem solving process which led to further recommendations on improving the work scheduling and delivery problems. The recommendations were never really installed, however, as people got back to their daily routines and forgot their agreements. The leadership group was unable, or unwilling, to require that the sales and production managers put the recommendations into effect.

This case involves a culture which is short on holding people accountable for solving problems and keeping their commitments to make performance improvements work. Human beings behave in the direction of their values, beliefs, and learned habits. But those values, beliefs and learned habits are influenced by the organizational culture in which they operate. When the culture does not emphasize accountability, even the most responsible employee finds it hard to remain motivated. It is difficult to work against a culture which lets others ignore accountability. When the culture is strong in recognizing and rewarding desired values, beliefs and habits, then employees either become acculturated in those directions, or withdraw and perhaps even leave the group. Ironically, a culture that is strongly dedicated to certain beliefs, values, and ways of doing things can often produce excellent performance results, but the ability to change the organization in a dramatically new direction is difficult because current patterns are so strongly reinforced.

2. The structure of the organization/team also influences the ability to put performance improvements into effect. The above case of the printing company is partly an issue of leadership and culture, but it is also partly a structural problem. The separation between production and sales leads to different agendas and constant conflict. Standard structural problems affecting performance and initiatives to improve performance include:

- Divisions between sales and production with no linkage or central authority resolving disputes
- Divisions between marketing and sales with those selling directly to the clients often having very different perspectives about what is needed in marketing strategies than those making marketing decisions

- Organizations with field-based personnel spread out across wide geographic areas having little support from or responsibility to either a regional or central authority

In these, as well as other structural scenarios, the performance problems usually come from a lack of a central authority with the dedication and persistence to improve performance and keep it at a high level. When people are separated by functional responsibility and reporting relationships (examples two and three above) or by geographic separation (example one just above) someone must be responsible and empowered to overcome those separations.

3. The location of the ultimate decision making power is also a key in making performance improvements permanent. If those ultimately responsible for performance in an organization/team are outside that group, then performance improvements always have to go through "higher level" approvals, and may well be vetoed by executives. This is the frustration of working in a large organizational structure where a relatively independent team is responsible for improving its results, but often does not have the final authority to make decisions that can produce the desired results.

D. Third factor affecting performance evaluation/ stabilization: Characteristics of the people involved

When the organization culture emphasizes results as well as commitment to people development, it will tend to attract people who are willing to take on challenges, grow, and learn. When the organizational culture has emphasized stability and routine, it will tend to attract and keep people who are much more comfortable with stability and often less willing to change as needed. The Missouri Division of Family Services case, discussed earlier in this chapter, is a classic example of the impact of group personality on the speed and ability to change for improved performance. Remember that the county welfare offices are newly responsible for labor exchange and finding jobs for welfare recipients as they move from welfare to work. This is very different from the determination of welfare eligibility and granting of benefits that they have done for many decades. But there are major struggles in making those changes. The initiative with Division of Family Services included a personality profile on all 250 participants for personal development. Out of four dimensions measured by this particular profile, the dimension measuring a tendency towards stability and steadiness was the primary dimension for close to 70% of the 250 welfare workers. It is not that they cannot change to meet the new services they are now required to provide, but they will change slowly and some will leave the organization out of frustration over unreasonable changes.

E. *Fourth factor affecting performance improvement evaluation/stabilization: Characteristics of the performance improvement intervention*

1. It is imperative that any performance improvement initiative maximize the opportunity for those in the organization/team to diagnose their own performance and build their own action plans for improvement. During diagnosis prior to the actual performance improvement intervention, the performance experts should have clearly identified performance gaps, causes, and perhaps even some potential solutions. But the work at improving performance has to include activity where those involved in the organization/team come to see the same issues identified in diagnosis, and refine and add to that diagnosis. Those who are to make performance improvements happen must also be involved in designing changes and actions for improvement. When outside consultants diagnose situations and develop performance improvement recommendations, the recommendations will tend to wither due to participant resistance. This is called the NIH syndrome, ("not invented here").

2. The performance improvement intervention must set clear goals for action with dates and responsibilities. This is easily done with strategic planning since setting objectives is a final step. But areas like improving management style, decreasing defects in the product, or building better customer service, may require effort at setting goals and deadlines. Still, the efforts are worth it.

3. Performance improvements must be measured and recognized when they occur. This is another place where the dedication of leadership at keeping performance improvements on center stage is essential.

4. The performance improvement initiative must include those in decision making authority over the organization/team needing development. This is often hard to do in a larger organization where plants or sales teams, for example, have a good deal of freedom and may want to work on their own performance, but do so independently of central executives. However, the leaders of these independent groups are well advised to keep higher level executives aware of what they are doing, even if those executives are not directly involved in the actual initiative.

F. *Conclusion*

The four factors identified above that affect the ability to evaluate and stabilize performance improvement are seldom all in perfect shape in any actual situation. But improvement in stabilizing performance improvement can also be achieved. For example:

- Leaders can be taught the importance of their role in making improvements permanent and can learn how to do it.
- Organizational cultures that do not emphasize goals, standards, measurement, and performance results can be changed through installation of goal setting systems and performance measurement.
- People who have personalities and learned habits that work against changing to improve performance can and do change. Even those whose behavior is highly habitized can learn new habits, though this, as well as culture change, takes time.
- There are many ways to design performance improvement interventions so as to maximize participant involvement in diagnosis of the issues and action planning for improvements.

It all takes dedication and persistence.

G. *Suggested action steps for organizational/team leaders*

1. Identify the areas where performance improvement efforts need to occur. What areas of your organization/team are essential for you to have high level performance to be successful?
2. Do you and your managers have sufficient goals and measurement to know if performance is sufficiently high? If so, do you have significant performance gaps? What are the causes of those gaps?
3. If you do not have sufficient goals, standards, and measures to know the performance level in essential areas of your organization, start developing those measures. Use concepts of goals, KPIs, and standards discussed here and in other referenced publications. Once you have a measure of performance success and gaps in important areas of your organization/team, go back to step two above, and then proceed on with the list below.
4. If significant performance gaps exist in your organization/team in areas that are essential, what performance interventions are available to effectively overcome those gaps? Stay with this process of decision making until the intervention is identified and put into effect. Make sure you or someone in your organization who is determined to see improvement stays with the project to initiate the performance improvement intervention as it unfolds.
5. Stay involved with the intervention sufficiently to learn what is being done and to show your commitment to that process. If there are areas such as knowledge and skills, attitude and motivation you need to improve your own performance, be fully involved in the intervention. Remember that your involvement will be a major factor in how others in your group see the importance of the initiative.
6. Evaluate the intervention. What worked and what did not? What needs to be kept and reinforced? How can you best do that? Look at

the rest of this chapter and the following chapter to get some ideas about reinforcing desirable performance improvement change.

7. If you are the leader of an independent team within a larger organization, work ways to keep higher level executives aware of what you are doing for performance improvement.

End notes

1. Cummings, Thomas G. and Worley, Christopher G., *Organizational Development and Change*, South-Western Publishing, Cincinnati, 1993, 480.

chapter fourteen

Comprehensive performance improvement

A. Introduction

The first chapter discussed the consequences of letting the wrong reasons such as current popularity of performance improvement initiatives motivate improvement efforts. This usually results in little or no diagnosis of real issues and therefore ineffective change programs. Stated more directly, the key stage in performance improvement is diagnosis. Conducting programs for reasons other than clearly identified performance gaps and improvement needs is wasteful, frustrating, and takes organizations, teams and people in the wrong direction.

Since diagnosis should be the driving force of performance improvement initiatives, then that diagnosis can also determine the scope of needed improvement. Which area of the organization(s) need to be improved? Sometimes leaders may be tempted to use a performance improvement initiative because the others doing it are team leaders who are co-workers. So, for example, the vice president for marketing decides to conduct a management development program and then the vice president for sales decides the same thing is needed in his/her department. Though familiarity with what the co-worker did in this scenario may be substantial, diagnosis may still be given short attention, and little of benefit will occur.

Sometimes, organizational/team leaders decide to develop their group in a number of ways rather than focusing just one area of performance deficiency. This is often the result of a sense leaders have that the organization/team is not performing at top level, even though data supporting that view is lacking. Here the performance vision and expectations of the organization/team leader(s) is paramount, rather than identified areas of performance deficiency. The following phrase is often heard: "We are good at what we do. We want to be even better." When leaders have this vision and expectations regarding performance, the need is to link a series of basic elements of high level performance.

B. Linking the basic elements of high level performance

There are a series of basic requirements to good organizational/team performance. These have been discussed throughout this book, using a one-at-a-time approach. Conducting one effort at performance improvement at a time is what most leaders choose. Few leaders have the ambition to take on a total campaign for performance improvement.

1. The first requirement for effective performance is a clearly stated strategic direction with strategic objectives or goals. In the case of a total organization such as a large corporation, a college, or a small business, this basic requirement is met through a strategic plan as discussed in Chapter 3. The same is true for an independent team, such as a large marketing/sales group with a good deal of discretion in how they approach their markets and which products they choose to offer to these markets.

A team that has little discretion in what products or services it provides, such as production or customer service groups, would find strategic planning for its group of little use, but direction is still needed for effective performance. This is where team goal setting, with goals cascading out of the strategy established further up the organizational structure, plays a critical role. Whatever the source of direction, strategic objectives and/or cascading goals serving strategy from a higher level, it is a must for effective performance.

2. The second requirement for effective performance is efficient technostructural relationships: that is the way the work and the reporting relationships are set up. These relationships should permit effective coordination between functions and business processes required for achieving the strategy or team goals. These chapters have not dealt extensively with restructuring the organization, largely because the best structure for any organization will vary with what kind of business they are in, and what kind of business processes are required. Therefore, there is no best structure for all or most organizations. However, any organizational/team structure should be based on the following guidelines:

- Reporting relationships and the division of work into groups should enhance work flow in areas like developing and producing products/services, delivering customer service, and enhancing sales and marketing results. Unfortunately, as demonstrated with a number of cases throughout this book, many structural relationships retard these processes.
- The organizational structure should include the ability to respond quickly to new opportunities or challenges. The term most often used in the literature is responsiveness or flexibility.

Restructuring is an effort at performance improvement which is overused partly because many leaders, believe they understand it and know how it is done. However, it is often done in ways producing frustration, bad morale, and unexpected consequences. The best approach is to use people familiar with the basic business processes to suggest the restructuring (product design and delivery, sales and marketing and customer service).

3. The third requirement for effective performance is to know the core competencies of the key positions in the organization/team. An overview of the meaning and role of core competencies in performance was provided in Chapter 10, "Performance Improvement at the Individual Level." As mentioned earlier, leaders interested in this performance improvement area have many sources of information available.[1] The point is that whatever method is used to define competencies, it is important to have a clear definition of underlying characteristics of an employee that are directly related to job performance.[2] This is the major flaw in most hiring processes; organizations seldom have clear and specific criteria of the competencies for performance success required for their most important jobs.

4. The fourth requirement for effective performance is a productive selection/hiring process (see Chapter 6). To be productive in this context means at least three things:

- Selection/hiring must take account of the strategic direction of the organization/team. For example, when an organization makes the strategic decision to enhance competitive advantage by aggressively creating new products and services to meet market needs, people with research and development experience in the new products areas are needed. People with experience at selling, servicing, or producing the new products may need to be hired. This is related to the principle that a specific strategy will impact all of the on-going processes of the organization.
- Selection/hiring should measure applicant possession of the core competencies connected to the positions being filled. This is true for job applicants from both the outside and the inside of the organization.
- As discussed in Chapter 6, selection/hiring should make use of all available information on the position and on the applicants.

5. The fifth requirement for effective organizational performance is to construct programs and methods for ongoing development of the knowledge and skills and motivation of the permanent teams. Those responsible for marketing, for example, must be tasked not only with meeting the basic goals of that function, but also meeting team development goals and individual team member performance and development goals. Failing that, knowledge and skills become rapidly outdated or rusty, motivation suffers, and attitudes and morale decline. Most teams and their members want to feel competent, be competitive, and do a good job. The leadership team, the

top team in the organization, should be no less responsible for ongoing development of their own knowledge and skills than the supervisors, customer service providers, production teams. The efforts at team development by leadership will be the model followed by other teams.

6. The sixth requirement for effective organizational performance is the existence of a system of performance management that on a daily basis reinforces the first five requirements listed above. This is where the dedication of even the most committed leadership can wane. As often as not, the resistance of leaders to a performance management system comes from their unwillingness to take the time to provide systematic feedback to their own direct reports. The importance of having constant reinforcement of the requirements for effective organizational performance is critical because the press of daily business makes those requirements difficult to maintain. Specifically, the performance management system should:

- Contain a section for the identification and measurement of the employee's responsibilities for the strategic objectives, as well as the operational goals and activities connected to their position.
- Contain a section for identification and evaluation of the employee's responsibilities for working with others in departments or teams separated from their own by the techno-structural design of the organization. In other words, part of the responsibilities of the employee (whatever their level of authority and responsibility) is to find ways not to let divisions into organizational units stop them from getting their work done and complete the various business processes in which they are involved. For example, if someone in contact with the customer is asked for a delivery date by that customer, contacting production to get the information should be a responsibility. If someone in marketing needs information from field sales, it is his responsibility to contact that group to get that data.

It is often surprising how, even in small organizations, separate silos of activity and responsibility created by the organizational structure and personality conflict, become a barrier to effective output. This, of course, is essentially what the Quality Movement of a few years ago was all about. While a good techno-structural design will be based on facilitating business processes, rather than retarding them, that is never enough. People make systems and structure work or not work. Attention to the daily performance of all employees is required. Leadership of the organization/team needs to empower their members to have access to others as required to get the work done. They should then make sure that empowerment is used.

- The performance management system must reinforce attention to the core competencies required for performance of the various jobs. The importance of this comes from a subtlety about core competencies often forgotten. That is that they are ideals which are only infrequently met totally. For example, if it becomes clear that a branch manager needs to show a great deal of patience and emotional maturity with her/his customers and direct reports because of the unusual stress in the branch, then that behavioral trait needs to be recognized in the feedback to branch managers, with specific examples of patience or lack of it. In a simpler situation, if the position requires a good deal of technical knowledge in the use of spreadsheets for data flow, then the performance management system should include attention to that skill as well as plans for developing it in employees when it is lacking.

- The performance management system should also reinforce the importance of attention to selection/hiring. If the person being managed and reviewed has a part in hiring, then goals and feedback regarding their effectiveness in that process need to be set and used. If the person has a part in orientation of new employees, then goals and Key Performance Indicators regarding that responsibility need to be set and feedback on performance in those areas should be provided in an ongoing way. The key point here is the following:

  "Organizations get what they pay attention to and measure."

- The performance management systems should also measure the success of efforts at building the performance of permanent teams. The leader of the team should set goals and get feedback regarding their own efforts at developing the group for which they have responsibility. The question sometimes arises regarding what should occur when the team leader is the CEO and has no boss for his or her own evaluation. The answer is simple. Who better to ask whether the performance of the team is being developed than the team members themselves?

7. The seventh requirement for effective organizational performance is the need for methods and systems for delivering training and development aimed at enhancing performance. In larger organizations, this should be a major focus of the human resources department and sometimes is. The key is to avoid what has been called "training for activity" and strive toward "training for impact." This is a brief reminder of the importance of performance diagnosis as the basis for performance improvement efforts, training and development or any other performance intervention.

8. The eighth requirement for effective organizational performance is the development of knowledge and skills for one of the most difficult managerial functions, coaching and counseling. This set of abilities is best developed as

a part of the Performance Management system, with managers trained to use coaching and counseling and employees educated to understand how to make the best use of the coaching and counseling they receive.

Three things are needed by those who coach skills or counsel in areas of attitude, motivation, etc. The first is the commitment to the importance of this role on the part of managers and more senior employees who can help with the development of new employees. The second is the comprehension of a series of concepts that help define the roles of the coach/counselor/mentor. The third, and perhaps more challenging need for effective coaching/counseling is the skills to do it. Somewhat surprisingly, many managers have never developed the ability to sit face to face with an employee and identify and discuss that employee's strengths and opportunities for improvement. This ability is, of course, much the same as what is needed in the evaluation step in performance management. The training and education of those providing coach/counseling can be understood in the context of the performance management system, where performance dimensions are identified to both the employee and the coach/counselor. It is always easier to discuss performance strengths and improvement needs when the performance dimensions have been agreed to by both parties.

A simple case will help clarify this point.

Organization: A consumer products company

- Products and Services
    food (for humans)
- Structure
    large field sales force
- Problem/Issues
    the business situation and the leadership style of the executive in charge of field sales and operations caused great difficulty for the CEO and the owner of the company. These two decision makers were confronted with the following dilemmas pertaining to the fields operations executive.

1. The business was undergoing rapid expansion focused primarily on entry into geographic markets were they had not been previously. The business was very successful financially and promised to be even more successful if the expansion into new markets was done well.
2. The field operations executive had decades of experience which made him invaluable in the projected entry into new markets.
3. The field operations executive had, however, developed a style of leadership that was seen as negative, overly controlling with his direct reports, and managed by identifying problems and going after and personally attacking people who he believed caused those problems.

As one of his direct reports stated, "The business is growing rapidly, we are succeeding in every way, and ... sees everything as a terrible crisis."

The CEO, supported by the owner, decided that the best course was to try to change the leadership style of the field operations executive, rather than lose him or continue to let him destroy morale. Turnover in field operations was higher than it needed to be and fear among field staff was at a high pitch.

The CEO used an approach of constant coaching and counseling, combined with outside leadership development programs aimed at changing the style of the operations executive. The CEO made it clear to all involved, including the executive's direct reports, that personal attacks were not going to be tolerated. He also used two other methods that substantially improved the situation. First, he commissioned an outside study of the morale of the field operations staff. Responses were confidential, and the attempt was to make clear to all involved the exact depth of morale and the specific issues involved. While the field operations executive did not completely accept the survey showing that morale was a major issue, nor his part in causing that low morale, he did begin to recognize that there was a problem which probably affected performance.

Secondly, the CEO moved the organization toward clear goal setting, especially in field operations. Sales goals, delivery goals, and goals pertaining to the amount of time product would be left at the customers' site were clearly established. This tended to increase the objectivity of the discussions between the field executive and his direct reports. Specific goals can take performance evaluation from vague and/or personal comments to somewhat more specific and measurable feedback.

The most recent description from the CEO in this coaching and counseling situation is that the field operations executive has significantly improved the objectivity of his feedback, has reduced the personal attacks, and has occasionally found the ability to provide sincere compliments for a job well done. While all is not perfect, and resentment from past attacks remains, performance remains high in the organization, and morale is measurably improved.

C. Conclusion

Linking the basic elements of high level performance into a comprehensive performance improvement effort takes enormous dedication and patience on the part of all those involved. Still, it can be done successfully. Those interested in reviewing actual cases of performance improvement efforts close to comprehensive in scope can review discussions throughout this book of Scottsdale Securities, Inc. and the hinge company.

D. Suggested action steps for organizational/team leaders

1. Review the eight requirements for effective performance and determine where you believe your organization/team meets the requirement and where you do not.
2. Be sure to involve others in your group in the process of assessing your organization/team.
3. Determine where the deficiencies are hurting your group significantly in achieving your vision of performance and the achievement of key goals.
4. In areas of poor performance because of a lack of requirements for effective performance, initiate discussion with your leadership team about needed actions.

End notes

1. Dubois, David D. Ph.D., *Competency-Based Performance Improvement*, HRD Press, Amherst, 1993.
2. Dubois, David D. Ph.D., *Competency-Based Performance Improvement*, HRD Press, Amherst, 1993, 9.
3. Robinson, Dana Gaines and Robinson, James C., *Training for Impact*, Josey-Bass Publishers, San Francisco, 1989.

chapter fifteen

Performance improvement efforts: Trade offs for leaders

A. Introduction

Organization/team leaders thinking about performance improvement face a profound dilemma. Are the benefits of performance improvement worth the costs of those efforts? Over the years, hundreds of leaders have been heard to express the following reasons, or variations thereof, for not engaging in performance improvement.

1. "We are doing well now. Why mess with success?"
2. "The benefits of performance improvement efforts are unpredictable. Why pay the costs for an unpredictable result?"
3. "Performance is hard to define. How can we know if initiatives will work?"
4. "We don't have time to work on improvement. We have to just do our work."

This general resistance to improvement efforts is the flip side of the "powerful need" for performance improvements discussed in Chapter 1. The need has resulted from intense competition, changes in technology and the desire to make use of it, wanting to feel competent, and similar socio/economic trends. Generalized felt need often leads to generalized felt rationalization about whether or not to actually get involved in specific improvement initiatives.

The key to resolving this dilemma for leaders is to get to specifics about performance and then decide if improvements are needed and are worth the costs in a specific organization. The greatest specificity about performance for any organization is a thorough strategic vision regarding that organization and the resulting goals from that strategic vision. Then diagnosis of performance gaps can occur, and decisions can be made about whether improvement needs to be initiated.

In an ideal way, performance improvement should be an ongoing effort, with all of those in an organization working to improve themselves and others for which they are responsible all the time. But realistically, daily operations take time, energy, and effort and "just getting the work done" sometimes seems to be all there is time and energy to do. However, because competition works to get ahead, products and services become outdated, technology changes rapidly, we often find that we cannot stand still. Deficiencies have crept up on us while we were busy engaged in daily operations.

When performance deficiencies occur, often as a result of the passing of time and the aggressiveness of competition, leaders need to decide on whether initiatives to improve performance are worth the effort and worry. The following template provides a general set of criteria for leaders to decide whether to pursue diagnosis of their specific organizational performance.

B. Performance status template: Organizational progress/systems

Program/System	Having It	Having Some of It	Not Having It
1. A written strategic plan	* Chance for shared effort	* Someone knows	* Production and marketing the same old way
A process for updating that plan	* Changes part of plan	* Changes driven by a few	* Changes poorly understood/ received
	* Keep current	* Sporadic updates	* Outdated
	*Production/selling focused	* Focus is spotty	* Coordination hard to achieve
2. Defined hiring process	* Potential use of a lot of information regarding applicants	* Can have varying success	* Random results
	Core competencies known	* Some clarity	* Hiring a burden
3. Attention to structure/ business process linkage	* Success in customer service efficiency	* Some areas productive	* Productivity blocked by structure
	* Restructuring with a plan	* Some needs not met	* Why do this?

Performance status template: Organizational progress/systems (continued)

Program/System	Having It	Having Some of It	Not Having It
4. Teams are consistently improved: subject matter and team functioning	* Professional * Pride in performance * Up-to-date	* Some individuals very good; some erratic	* Teams in name only
5. Performance management system with goals and KPIs	* We all know performance level	* Some manager/ employees know	* Feedback spotty and subjective
6. Consistent education/ training and development	* Learning is valued	* Good and then not good	* Same old job * Wait until the weekend
Linked to strategy Based on diagnosed need	* The organization knows/cares	* Occasional Escape	* Years to retirement?
Involves knowledge and skills for total life	* Ongoing	* A few have it	* Just a job
7. Coaching/ counseling practiced	* Improving	* Some being helped	* All on their own

C. Conclusion and final suggested action steps for organizational/team leaders

Any template such as the one above is an approximation of consequences from having or not having organizational elements enhancing performance. It is based on probabilities that certain program elements will have predictable results and lack of those program elements will also have predictable results. Those results are about performance.

The template is not a substitute for individual leaders assessing their own organization/team performance. It is a starting point for consideration.

Step 1. Review the above template and detail your organization/ teams' status in regard to each of the program/system elements.
Step 2. Discuss your perceptions regarding performance status with your trusted and knowledgeable co-workers.
Step 3. What next?

When organizational/team leaders believe their unit can perform significantly better, the above template can be a very useful starting point for action. It can help leaders know what programs and systems they have in place and what is missing.

Index